Computer Simulation
A Foundational Approach Using Python

CHAPMAN & HALL/CRC
COMPUTER and INFORMATION SCIENCE SERIES

Series Editor: Sartaj Sahni

PUBLISHED TITLES

ADVERSARIAL REASONING: COMPUTATIONAL APPROACHES TO READING THE OPPONENT'S MIND
Alexander Kott and William M. McEneaney

COMPUTER-AIDED GRAPHING AND SIMULATION TOOLS FOR AUTOCAD USERS
P. A. Simionescu

COMPUTER SIMULATION: A FOUNDATIONAL APPROACH USING PYTHON
Yahya E. Osais

DELAUNAY MESH GENERATION
Siu-Wing Cheng, Tamal Krishna Dey, and Jonathan Richard Shewchuk

DISTRIBUTED SENSOR NETWORKS, SECOND EDITION
S. Sitharama Iyengar and Richard R. Brooks

DISTRIBUTED SYSTEMS: AN ALGORITHMIC APPROACH, SECOND EDITION
Sukumar Ghosh

ENERGY-AWARE MEMORY MANAGEMENT FOR EMBEDDED MULTIMEDIA SYSTEMS:
A COMPUTER-AIDED DESIGN APPROACH
Florin Balasa and Dhiraj K. Pradhan

ENERGY EFFICIENT HARDWARE-SOFTWARE CO-SYNTHESIS USING RECONFIGURABLE HARDWARE
Jingzhao Ou and Viktor K. Prasanna

EVOLUTIONARY MULTI-OBJECTIVE SYSTEM DESIGN: THEORY AND APPLICATIONS
Nadia Nedjah, Luiza De Macedo Mourelle, and Heitor Silverio Lopes

FROM ACTION SYSTEMS TO DISTRIBUTED SYSTEMS: THE REFINEMENT APPROACH
Luigia Petre and Emil Sekerinski

FROM INTERNET OF THINGS TO SMART CITIES: ENABLING TECHNOLOGIES
Hongjian Sun, Chao Wang, and Bashar I. Ahmad

FUNDAMENTALS OF NATURAL COMPUTING: BASIC CONCEPTS, ALGORITHMS, AND APPLICATIONS
Leandro Nunes de Castro

HANDBOOK OF ALGORITHMS FOR WIRELESS NETWORKING AND MOBILE COMPUTING
Azzedine Boukerche

PUBLISHED TITLES CONTINUED

HANDBOOK OF APPROXIMATION ALGORITHMS AND METAHEURISTICS
Teofilo F. Gonzalez

HANDBOOK OF BIOINSPIRED ALGORITHMS AND APPLICATIONS
Stephan Olariu and Albert Y. Zomaya

HANDBOOK OF COMPUTATIONAL MOLECULAR BIOLOGY
Srinivas Aluru

HANDBOOK OF DATA STRUCTURES AND APPLICATIONS
Dinesh P. Mehta and Sartaj Sahni

HANDBOOK OF DYNAMIC SYSTEM MODELING
Paul A. Fishwick

HANDBOOK OF ENERGY-AWARE AND GREEN COMPUTING
Ishfaq Ahmad and Sanjay Ranka

HANDBOOK OF GRAPH THEORY, COMBINATORIAL OPTIMIZATION, AND ALGORITHMS
Krishnaiyan "KT" Thulasiraman, Subramanian Arumugam, Andreas Brandstädt, and Takao Nishizeki

HANDBOOK OF PARALLEL COMPUTING: MODELS, ALGORITHMS AND APPLICATIONS
Sanguthevar Rajasekaran and John Reif

HANDBOOK OF REAL-TIME AND EMBEDDED SYSTEMS
Insup Lee, Joseph Y-T. Leung, and Sang H. Son

HANDBOOK OF SCHEDULING: ALGORITHMS, MODELS, AND PERFORMANCE ANALYSIS
Joseph Y.-T. Leung

HIGH PERFORMANCE COMPUTING IN REMOTE SENSING
Antonio J. Plaza and Chein-I Chang

HUMAN ACTIVITY RECOGNITION: USING WEARABLE SENSORS AND SMARTPHONES
Miguel A. Labrador and Oscar D. Lara Yejas

IMPROVING THE PERFORMANCE OF WIRELESS LANs: A PRACTICAL GUIDE
Nurul Sarkar

INTEGRATION OF SERVICES INTO WORKFLOW APPLICATIONS
Paweł Czarnul

INTRODUCTION TO NETWORK SECURITY
Douglas Jacobson

LOCATION-BASED INFORMATION SYSTEMS: DEVELOPING REAL-TIME TRACKING APPLICATIONS
Miguel A. Labrador, Alfredo J. Pérez, and Pedro M. Wightman

PUBLISHED TITLES CONTINUED

Computer Simulation

A Foundational Approach Using Python

Yahya E. Osais

CRC Press is an imprint of the
Taylor & Francis Group, an **informa** business

A CHAPMAN & HALL BOOK

CRC Press
Taylor & Francis Group
6000 Broken Sound Parkway NW, Suite 300
Boca Raton, FL 33487-2742

Printed on acid-free paper
Version Date: 20171024

International Standard Book Number-13: 978-1-498-72682-5 (Hardback)

Visit the Taylor & Francis Web site at
http://www.taylorandfrancis.com

and the CRC Press Web site at
http://www.crcpress.com

To my wife, Asmahan,
and my daughters, Renad, Retal, and Remas.

Contents

PART III Problem-Solving

Part V Case Studies

List of Programs

List of Figures

List of Tables

Foreword

Computer simulation is an effective and popular universal tool. It can be applied to almost all disciplines. Typically, simulation involves two key steps: modeling and implementation, which are highlighted in this book. Modeling can performed using event graphs, which are revived by this book. As for implementation, complete Python programs are given, which is considered a new effort. This is an interesting combination as the translation process from models to programs is straightforward.

The book also dedicates a complete chapter on the popular Monte Carlo simulation method. This chapter covers several variance-reduction techniques along with their implementation in Python. Three interesting case studies are discussed in detail. The book features good examples and exercises for readers and students.

This book is highly recommended for a graduate course in modeling and simulation. It is also recommended for an introductory course in modeling and simulation for a senior undergraduate course. In addition, it can be a good reference for researchers and working engineers and scientists who work on modeling and simulation and optimization. This book is a good addition to the field of modeling and simulation. I hope you will enjoy the book as much as I have enjoyed reviewing it.

Mohammad S. Obaidat, Fellow of IEEE and Fellow of SCS
Past President of the Society for Modeling and Simulation International, SCS
Founding Editor in Chief, Security and Privacy Journal, Wiley
Editor-in-Chief, International Journal of Communication Systems
June 2017

Preface

This book is not just another book on discrete-event simulation. It emphasizes modeling and programming without sacrificing mathematical rigor. The book will be of great interest to senior undergraduate and starting graduate students in the fields of computer science and engineering and industrial engineering. The book is designed for a one-semester graduate course on computer simulation. Each chapter can be covered in about one week. The instructor is also encouraged to dedicate one week for learning the Python programming language. Appendix A can be used for this purpose. A basic knowledge of programming, mathematics, statistics, and probability theory is required to understand this book.

The book has the following features. First, a simulation program is clearly divided into two parts: simulator and model. In this way, implementation details based on a specific programming language will not coexist with the modeling techniques in the same chapter. As a result, student confusion is minimized. The second feature of the book is the use of the Python programming language. Python is becoming the tool of choice for scientists and engineers due to its short learning curve and many open-source libraries. In addition, Python has a REPL[1] which makes experimentation much faster. The third feature is the use of event graphs for building simulation models. This formalism will aid students in mastering the important skill of simulation modeling. A complete chapter is dedicated to it. The book also features a complete chapter on the Monte Carlo method and variance-reduction techniques. Several examples are given along with complete programs.

The book is divided into four parts. The first part represents a complete course on the fundamentals of discrete-event simulation. It is comprised of chapters 1 to 6. This first part is appropriate for an undergraduate course on discrete-event simulation. Materials from other chapters can be added to this course. For example, chapter 10 and 11 should be covered in full if time permits. For an advanced course on computer simulation, the second and third part should be fully covered. The case studies in the fourth part can be covered if time permits. In such a course, the emphasis should be on model building and programming.

---

[1]REPL = Read-Evaluate-Print Loop

To the Reader

While writing this book, I had assumed that nothing is obvious. Hence, all the necessary details that you may need are included in the book. However, you can always skip ahead and return to what you skip if something is not clear. Also, note that throughout this book, "he" is used to to refer to both genders. I find the use of "he or she" disruptive and awkward. Finally, the source code is deliberately inefficient and serves only as an illustration of the mathematical calculation. Use it at your own risk.

Website

The author maintains a website for the book. The address is `http://faculty.kfupm.edu.sa/coe/yosais/simbook`. Presentations, programs, and other materials can be downloaded from this website. A code repository is also available on Github at `https://github.com/yosais/Computer-Simulation-Book`.

Acknowledgments

I would like to thank all the graduate students who took the course with me while developing the material of the book between 2012 and 2017. Their understanding and enthusiasm were very helpful.

I would also like to thank King Fahd University of Petroleum and Minerals (KFUPM) for financially supporting the writing of this book through project number **BW151001**.

Last but not least, I would like to thank my wife for her understanding and extra patience.

Yahya Osais
Dhahran, Saudi Arabia
2017

About the Author

Yahya E. Osais is a faculty member in the department of computer engineering at King Fahd University of Petroleum and Minerals (KFUPM), Dhahran, Saudi Arabia. He earned his B.Sc. and M.Sc. from the same department in 2000 and 2003, respectively. In 2010, he obtained his Ph.D. from the department of systems and computer engineering at Carleton University, Ontario, Canada.

Dr. Osais regularly teaches a graduate course in computer simulation for students in the college of computer science and engineering at KFUPM. He also teaches courses on computer engineering design and web technologies. His current research interest includes stochastic modeling and simulation, cyber-physical systems, and the Internet of things.

Abbreviations

RV	Random Variable
CDF	Cumulative Distribution Function
iCDF	Inverse CDF
PDF	Probability Distribution Function
PMF	Probability Mass Function
BD	Birth-Death
LFSR	Linear Feedback Shift Registers
RNG	Random Number Generator
RVG	Random Variate Generator
REG	Random Event Generator
IID	Independent and Identically Distributed

Symbols

Variable that are used only in specific chapters are explained directly at their occurrence and are not mentioned here.

λ	Average Arrival Rate	*Pronounced as "lambda"*
μ	Average Service Rate	*Pronounced as "meu"*
IAT	Average Inter-Arrival Time between any Two Consecutive Packets	
ST	Average Service Time of a Packet	$IAT = \frac{1}{\lambda}$ $ST = \frac{1}{\mu}$
T	Total Simulation Time	
$clock$	Current Simulation Time	
IAT_i	Time until the arrival of Packet i	
ST_i	Service Time of Packet i	
A_i	Arrival Time of Packet i	$A_i = clock + IAT_i$
D_i	Departure Time of Packet i	$D_i = clock + ST_i$
μ	Population Mean	
σ^2	Population Variance	
σ	Population Standard Deviation	
$\bar{x}$	Sample Mean	
s^2	Sample Variance	
s	Sample Standard Deviation	

I

The Fundamentals

## CHAPTER 1

Introduction

"The purpose of computing is insight, not numbers."
−Richard Hamming

The purpose of this chapter is to motivate the importance of simulation as a scientific tool. The chapter also introduces some essential concepts which are needed in the rest of the book. The lifecycle of a simulation study is described here along with an example. In addition, the advantages and limitations of simulation are discussed. The reader is urged to carefully read this chapter before moving on to the next ones.

1.1 THE PILLARS OF SCIENCE AND ENGINEERING

Science and engineering are based on three pillars: observation, experimentation, and computation. Figure 1.1 uses the analogy of a table with three legs to show the relationship between these three tools and science and engineering. Historically, humans have been using observation and experimentation to acquire new knowledge (i.e., science) and then apply the newly acquired knowledge to solve problems (i.e., engineering). This approach is very effective because the actual phenomenon (system) is observed (utilized). However, as the complexity increases, observation and experimentation become very costly and cumbersome. This is when computation becomes the only tool that can be used.

The outcome of an observational study is a set of facts. For example, if a burning candle is covered with a glass cup, it will eventually go out on its own. This is the observation. Scientists had to do research before they could realize the reason for this phenomenon. The reason is that there is still oxygen inside the glass cup which will eventually be used up by the flame. Once all the oxygen is consumed, the candle goes out.

On the other hand, experimentation is the act of making an experiment.

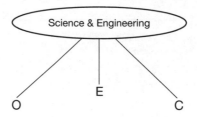

Figure 1.1
The three pillars of science and engineering: Observation (O), Experimentation (E), and Computation (C). By analogy, the table needs the three legs to stay up.

An experiment is a physical setup. It is performed to make measurements. Measurements are raw data. Experimentation is popular among scientists.

The output of the system is recorded as it occurs in an observational study. Furthermore, the response of the system is not influenced in any way and the environment in which the system operates cannot be manipulated. In experimentation, however, we can manipulate the environment in which the system operates and influence the response of the system.

A computation is a representation of the phenomenon or system under study in the form of a computer program. This representation can be as simple as a single mathematical equation or as complex as a program with a million lines of code. For mathematical equations, there are tools like calculus and queueing theory that can be used to obtain closed-form solutions. If a closed-form solution, on the other hand, cannot be obtained, approximation techniques can be used. If even an approximate solution cannot be obtained analytically, then computation has to be used.

In this book, we are interested in the use of computation as a tool for understanding the behavior of systems under different conditions. This goal is achieved by generating time-stamped data which is then statistically analyzed to produce performance summaries, like means and variances. The type of computation performed by the program which generates this type of data is referred to as event-oriented simulation. Developing such simulation programs is an art. The good news is that you can acquire this skill by practice. Therefore, it is recommended that you carefully study the examples in the book.

1.2 STUDYING THE QUEUEING PHENOMENON

Consider the situation in Figure 1.2 where five people have to wait in a *queue* at the checkout counter in a supermarket. This situation arises because there is only one cashier and more than one person wants to have access to him. This phenomenon is referred to as *queueing*. Let us see how observation, experimentation, and computation can be used to study this phenomenon.

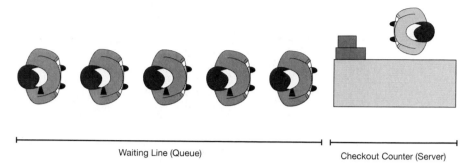

Figure 1.2

A queue at a checkout counter in a supermarket. A phenomenon arising whenever there is a shared resource (i.e., the cashier) and multiple users (i.e., the shoppers).

If we want, for example, to estimate the average time a customer spends at the checkout counter, we should manually record the time each customer spends waiting to be served plus the service time. Therefore, for each customer, we have to keep track of two times: (1) time customer joins the queue (*arrival time*) and (2) time customer leaves the system (*departure time*) . As shown in Figure 1.2, the system is made up of the waiting line and checkout counter. Clearly, performing this observational study is costly and cumbersome.

For example, we can control the maximum number of people who should wait in the queue to shorten the time of the experiment. We can also introduce a policy that only customers with no more than seven items can join the queue. We can also ask customers to help us in collecting the necessary data. For example, each customer can be asked to record the times at which he joins and leaves the queue. This will surely reduce the cost of the experiment.

A study based solely on computation (i.e., simulation) is significantly less costly. It requires only developing a computer program based on a sound model of the situation under study. A computational approach is also the most flexible one since it gives us a full control over both the environment and system.

1.3   WHAT IS SIMULATION?

Simulation is the process of representing a system[1] by a model and then executing this model to generate raw data. The raw data is not useful by itself. It must be statistically processed to produce insights about the performance of the system. These four activities are represented by four gray boxes in Figure 1.4. They are part of a larger framework of activities that constitute

---

[1]From now on, we are going to use the word "system" as a synonym for the words "phenomenon" and "situation."

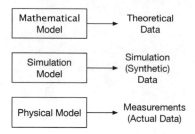

Figure 1.3
Types of models and the data generated from them.

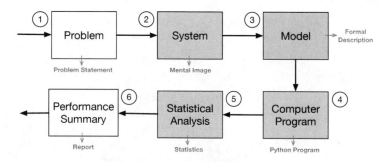

Figure 1.4
Phases of a simulation study.

the elements of a simulation study. More about this will be said in the next section.

The model is a conceptual representation of the system. It represents a modeler's understanding of the system and how it works. A computer is used to execute the model. Therefore, the model must first be translated into a computer program using a programming language like Python.[2] The execution of the computer program results in the raw data.

The raw data is also referred to as simulation data. It is synthetic because it is not actual data. Actual data is collected from a physical model of the system. There is another type of data called theoretical data which is generated from a mathematical model of the system. Figure 1.3 shows these types of models and the types of data generated from them.

1.4 LIFECYCLE OF A SIMULATION STUDY

Simulation is just a tool. When you study a system using simulation, you conduct a simulation study. As Figure 1.4 shows, there are six phases in a simulation study. In the first phase, a detailed description of the problem you

---

[2]https://www.python.org.

want to solve using simulation is developed. A good problem description helps in defining the boundaries of the system, scope of the study, and guide the model development. The problem description should include a set of questions and objectives. Also, it should include a description of the assumptions and performance metrics which will be used in making decisions in the last phase of the study. The raw data which is needed to compute the performance metrics must also be clearly defined. A more formal name for the outcome of this phase is problem statement.

The system and environment in which it operates should be described in the second phase. Only the details which are relevant to the simulation study should be captured. The skill of abstraction, which is discussed in Chapter 2, becomes very handy in this stage. The outcome of this phase is a mental image of the system and its operation. A mental image of a system represents one's understanding of how the system works.

The third phase is about developing a conceptual model of the system. Using elementary concepts, which is covered in Chapter 2 and the formal language of event graphs covered in Chapter 6, a model of the system can be developed. Modeling is still an art and not a science. Thus, it is recommended that you carry out this phase iteratively. That is, you start with a simple model and you keep improving it until you are confident that your model captures the actual system you intend to study. The outcome of this phase is a formal description of the system.

In the fourth phase, the model is encoded in a computer language. In this book, we use the Python programming language for this purpose. Translating a model into a computer program is not an easy task. You also need to write your code in a certain way to ease collection of the data necessary for the next phase. In order to succeed in this phase, you need to familiarize yourself with the programming language and its capabilities. In this book, we cover Python in sufficient depth. Many implementation examples are given throughout the book. The outcome of this phase is a Python implementation of the model.

The result of executing the model is a set of raw data which should be statistically analyzed. Statistics like the mean, variance, and confidence intervals should be derived in this phase. These statistics are only approximations of the values of the performance metrics stated in the problem statement. Their reliability depends on quality and size of the raw data collected in the previous phase. You may need to execute the model several times and/or for a longer time to get a good data set. You will also have to make sure your data set is *IID*. These details will be discussed further in Chapter 11. The outcome of this phase is a set of statistics which summarize the performance of the system.

Finally, in the last phase, a summary in the form of a report must be prepared. This report should include answers for the questions stated in the problem statement. The simulation study is a failure if the report does not achieve the intended objectives declared in the problem statement. The report should also include a set of conclusions and recommendations based on the

Table 1.1
Description of the phases of a simulation study of the system in Figure 1.2.

No.	Phase	Description
1	Problem	Customers experience delays longer than 5 minutes. The checkout process has to be speeded up. Potential solutions include changing the cashier and installing a new software system. The raw data to be collected include the delay experienced by each customer i (D_i) which is defined as the difference between his departure time (D_i) and arrival time (A_i).
2	System	Customers, waiting line, and cashier.
3	Model	A customer arrives at the system. If the cashier is free, he will be served immediately. Otherwise, he has to wait. Service time of each customer is random.
4	Computer Program	Model is expressed in Python code.
5	Statistical Analysis	Response time of the system (i.e., the average delay). $T_{avg} = \frac{\sum_i^N D_i}{N}$, where N is the number of participating customers.
6	Performance Summary	Response time for each possible solution. Pick the one that gives the best response time as the optimal solution.

statistical analysis of the simulation data. A conclusion which sums up the evidence for the decision maker must be given. The summary report is the outcome of this phase and the overall study.

Table 1.1 contains the details of a simulation study of the situation in Figure 1.2. The goal of the study is to find the best solution from a set of suggested solutions for speeding up the checkout process. The system has been defined and a conceptual model has been developed as explained in the table. Only one performance metric is used in this study, which is the average delay through the system. The raw data needed to compute this metric has been decided in the problem statement. The performance summary will include the delay introduced by each possible solution. The management will pick the one that causes the least delay if it can be afforded, of course.

1.5   ADVANTAGES AND LIMITATIONS OF SIMULATION

As the complexity of the system under study increases, analytical tools may fail to capture the details that are of interest to the modeler. Even for simple systems, simulation can sometimes provide very insightful information. The following are some of the reasons why we need simulation.

1. There is no need to build the physical system under study and then observe it. Thus, knowledge about the behavior of the system can be acquired with a minimum cost.

2. Critical scenarios can be investigated through simulation with less cost and no risk.

3. Using a simulation model, the effect of changing values of system variables can be studied with no interruption to the physical system.

4. Simulation is more flexible and convenient than mathematical analysis. Also, the modeler can avoid the hassle of dealing with mathematical equations.

5. In simulation, there is no need for simplifying assumptions like in mathematical models where such assumptions are needed to make the models tractable.

6. Simulation allows us to compress and expand the behavior of the system under study. For example, several years' worth of system evolution can be studied in a few minutes of computer time. Also, one second of simulation time can be expanded to several hours of computer time.

Like any other tool, simulation has limitations. The following are some of them:

1. The outcome of a simulation study is an estimate subject to a statistical error. For example, different simulation runs typically produce different numbers although the same simulation model is used.

2. Simulation can become costly and time consuming. For example, very powerful computers and skillful people are required.

3. Simulation models are not easy to develop. Existing methodologies are not universal. This is why development of simulation models is still an art, not a science.

4. Existing programming languages are not designed to support simulation. Thus, a lot of programming is involved.

1.6 OVERVIEW OF THE BOOK

The rest of the book is structured as follows:

Chapter 2 is about building conceptual models. These conceptual models are what we call simulation models. Also, in this chapter, state diagrams are introduced as a tool for describing the dynamic behavior of systems. The main example used in this chapter is the single-server queueing system, which also serves as our running example throughout the book.

Chapter 3 is a review of probability using a computational approach. In this chapter, the reader is exposed to the Python programming language for the first time in the book. Therefore, the reader is strongly encouraged to go through Appendix A thoroughly.

Chapter 4 is a review of random variables and stochastic processes using also a computational approach. In this chapter, the queueing phenomenon is discussed again. Also, the notion of a state space of a dynamic system is explained.

Chapter 5 discusses the simplest queueing system which is the single-server, single-queue system. In this chapter, some basic performance laws are introduced. Manual simulation is also covered in this chapter.

Chapter 6 is about the collection and statistical analysis of data that results from executing the simulation model. The notion of an output variable as a mechanism for collecting data is introduced in this chapter. In addition, all the necessary statistical concepts such as point estimates and confidence intervals are discussed in sufficient detail. The method of independent replications and how to deal with the bias due to the warm-up period are also discussed.

Chapter 7 is about modeling using event graphs. This is a very important intermediate step that helps the reader to develop his modeling skills. An event graph shows how events interact inside a simulation model.

Chapter 8 explains the difference between time-driven and event-driven simulation. It also describes in detail how an event-driven simulation program is constructed. All the necessary concepts and language features are covered. Complete programs are shown and discussed in depth.

Chapter 9 covers the Monte Carlo method. This method is used for solving problems that do not require a full-blown simulation model. Diverse examples are used to demonstrate the practicality of the method. Further, the notion of variance reduction is introduced and several techniques are discussed.

Chapter 10 is about generating random numbers from non-uniform probability distributions. Such numbers are referred to as random variates. These numbers are used to represent the lifetimes of random phenomena that occur inside the simulation model. Examples of such random phenomena are the time until the next failure or departure.

Chapter 11 is about generating random numbers from a uniform probability distribution over the interval (0, 1). This procedure is the source of randomness in the simulation program. It drives the process of generating random variates. Several tests for randomness are covered to ensure the quality of the generated random numbers.

Chapter 12 contains several case studies. The purpose of these case studies is to show how the concepts and skills explained throughout the book can be applied. Each case study represents a complete simulation study.

Four appendices are added to complement the core material of the book. **Appendix A** serves as an introduction to the Python programming language. **Appendix B** describes an object-oriented framework for discrete-event simulation. The object-oriented paradigm is very popular among software developers. This is because it enables code reuse and easy code maintenance. Finally, **appendices C** and **D** both contain standard statistical tables.

1.7 SUMMARY

Simulation is a tool that can be used for performing scientific studies. It may not be the first choice. But, it is definitely the last resort if a physical or mathematical model of the system under study cannot be constructed. The main challenge in simulation is developing a sound model of the system and translating this model to an efficient computer program. In this book, you will learn the skills that will help you to overcome this challenge.

## CHAPTER 2

Building Conceptual Models

"Modeling means the process of organizing knowledge about a given system."
—Bernard Zeigler

This chapter is about building conceptual models. It describes the transition from a mental image of the system to a conceptual model that captures the structure of the system and its behavior. The single-server queueing system is formally introduced in this chapter. It serves as our main example throughout the book. State diagrams are also introduced in this chapter. They are used to show how the behavior of a system which is expressed as changes in the values of state variables evolves as a result of the occurrence of events.

2.1   WHAT IS A CONCEPTUAL MODEL?

A conceptual model is a representation of a system using specialized concepts and terms. For example, a mathematical model of a system can be thought of as a conceptual model that is constructed using specialized concepts such as constants, variables, and functions and specialized terms such as derivative and integral. However, before a conceptual model can be built, a mental image of the system under study must be developed in the mind of the modeler. As shown in Figure 2.1, having a good mental image of the system is the first step towards a good conceptual model. A mental image reflects how the modeler perceives the system and its operation. The mental image should include only those aspects of the system that are necessary for the simulation study.

Figure 2.2 tells us that different mental images can be developed for the same system. They differ only in their amount of details. Typically, the first mental image (i.e., level 1) is the simplest one. As you add more details, they

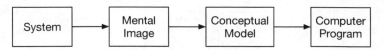

Figure 2.1
A mental image of the system and its behavior must be developed before a conceptual model can be constructed.

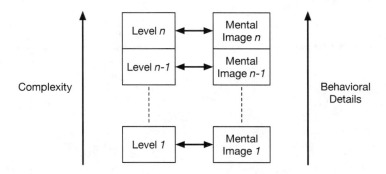

Figure 2.2
Different mental images can be developed for the same system. They include different levels of details. Complexity increases as you add more details.

become more complex. Eventually, the mental image cannot fit in the head of the modeler. So, he needs other tools to manage them.

The type of systems we deal with in this book is called Discrete-Event Systems (DESs). A DES is made up of elements referred to as entities. For example, in the supermarket example in Chapter 1 (see Figure 1.2), the entities are the customers, cashier, and queue. Entities participate in activities which are initiated and terminated by the occurrence of events. Events are a fundamental concept in simulation. They occur at discrete points of time. The occurrence of an event may cause other events to occur. Entities change their state upon occurrence of events. For example, the cashier becomes busy when a customer starts the checkout process. More will be said about these concepts in the next section. For now, you just need to become aware of them.

What is in a Name?

The meaning of the name "discrete-event simulation" is not clear for many people. The word "event" indicates that the simulation is advanced by the occurrence of events. This is why the name "event-driven simulation" is also used. The word "discrete" means that events occur at discrete points of time. So, when an event occurs, the simulation time is advanced to the time at which the event occurs. Hence, although time is a continuous quantity in reality, simulation time is discrete.

2.2 ELEMENTS OF A CONCEPTUAL MODEL

A conceptual model can be constructed using five elements:

1. Entity,

2. Attribute,

3. State Variable,

4. Event, and

5. Activity.

Next, each one of these elements is discussed in detail. They will be used in the next section to build a conceptual model for the single-server queueing system.

2.2.1 Entities

An entity represents a physical (or logical) object in your system that must be explicitly captured in the model in order to be able to describe the overall operation of the system. For example, in Figure 2.6, in order to describe the depicted situation, an entity whose name is coffee machine must be explicitly defined. The time the coffee machine takes to dispense coffee contributes to the overall delay experienced by people. The coffee machine is a *static* entity because it does not move in the system and its purpose is to provide service only for other entities.

A person is another type of entity that must be defined in the model. A person is a *dynamic* entity because it moves through the system. A person enters the system, waits for its turn, and finally leaves the system after getting his coffee.

A static entity maintains a state that can change during the lifetime of the system. On the other hand, dynamic entities do not maintain any state. A dynamic entity typically has attributes which are used for storing data.

2.2.2 Attributes

An entity is characterized using attributes, which are local variables defined inside the entity. For example, a person can have an attribute for storing the time of his arrival into the system (i.e., arrival time). Another attribute can be defined to store the time at which the person leaves the system (i.e., departure time). In this way, the time a person spends in the system is the difference between the values stored in these two attributes.

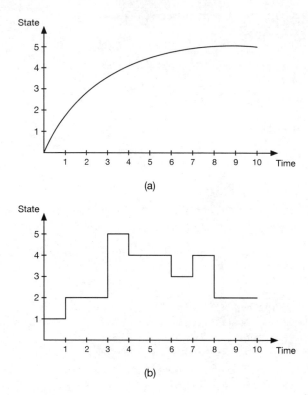

Figure 2.3
A continuous state variable takes values from a continuous set (e.g., [0, 5] in (a)). A discrete state variable, on the other hand, takes values from a discrete set (e.g., {0, 1, 2, 3, 4, 5} in (b)).

2.2.3 State Variables

The state of the system under study is represented by a set of *state variables*. A state variable is used to track a property of a static entity over time. For example, for a memory module in a system, its state could be the number of data units it currently stores. Another example is the state of a cashier in the supermarket example. It is either free or busy.

A state variable is said to be *continuous* if it takes values that change continuously over time. However, if the value of a state variable is from a discrete set, then it is referred to as a *discrete* state variable. Figure 2.3 illustrates the difference between these two types of state variables. Note that the value of a continuous state variable changes with every change in time. The value of the discrete state variable, however, changes at discrete points of time.

A state variable changes its value when an *event* occurs inside the system. An event acts as a stimulus for the system to change its state. For example, when you start your car, you actually generate an event that stimulates the

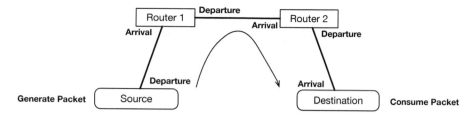

Figure 2.4
Events are used to move dynamic entities through a system. A packet is moved
from a source to a destination through two routers using eight events.

car and causes it to turn itself on. Because of this *startup* event, the state of
the car has changed from *OFF* to *ON*. Hence, an event triggers a change in
one or more state variables in a conceptual model.

2.2.4   Events

An event represents the occurrence of something interesting inside the system.
It is a stimulus that causes the system to change its state. For instance, in
the supermarket example, the arrival of a new customer represents an event
which will cause the state variable representing the number of people waiting
in line to increase by one. The departure of a customer will cause the cashier
to become free.

Events can also be used to delimit activities and move active entities in
the system. For example, in Figure 2.4, a packet is moved from a source to
a destination using eight events. The first event generates the packet. After
that, a sequence of *departure* and *arrival* events moves the packets through the
different static entities along the path between the source and destination. For
instance, for *router 1*, its arrival event indicates that the packet has arrived
at the router and it is ready for processing. After some delay, the same packet
leaves the router as a result of a departure event.

2.2.5   Activities

An activity is an action which is performed by the system for a finite (but
random) duration of time. As shown in Figure 2.5, an activity is delimited
by two distinct events. The initiating event starts the activity. The end of
the activity is scheduled at the time of occurrence of the terminating event.
The difference in time between the two events represents the duration of the
activity.

In the supermarket example, an important activity is the time a customer
spends at the checkout counter. The duration of this activity depends on how
many items the customer has. Durations of activities are modeled as random
variables. Random variables are covered in Chapter 4.

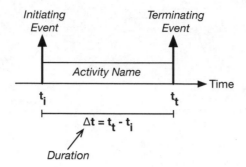

Figure 2.5
An activity is delimited by two events and lasts for a random duration of time.

2.3 THE SINGLE-SERVER QUEUEING SYSTEM

Consider the situation depicted in Figure 2.6 where there is one coffee machine and multiple users. Only one user can use the machine at a time. Thus, the others have to wait in a queue. This phenomenon is referred to as *queueing*. It is very common in places like supermarkets and airports. It also arises in computer systems and networks. For example, inside a router, packets destined to the same output port have to wait for their turns.

Before we can develop a conceptual model for the queueing situation in Figure 2.6, we have to first develop a mental image of it. A simple mental image could be the following:

Every day at 8 am, after check-in, each person goes directly to the kitchen to get his coffee. There is only one coffee machine in the kitchen. As a result, if someone already is using the machine, others have to wait. People use the machine in the order in which they arrive. On average, a person waits for a non-zero amount of time before he can use the machine. This amount of time is referred to as the delay.

Table 2.1 shows the details of the conceptual model which results from the above mental image. The same information is presented pictorially in Figure 2.7. There are three entities. The queue and server are static entities. A person is a dynamic entity since he can move through the system. Three events can be defined: (1) arrival of a person into the system, (2) start of service for a person, and (3) departure of a person from the system. Remember that these events are used to move the person entity through the system.

Two state variables need to be defined to keep track of the number of persons in the queue and the state of the server. Everytime a person arrives into the system, the state variable of the server, S, is checked. If its value is Free, it means the server can serve the arriving person. On the other hand, if the value is Busy, the arriving person has to join the queue and wait for his

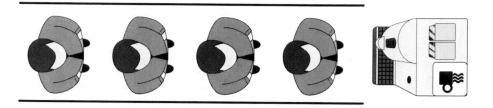

Figure 2.6
A queueing phenomenon emerges whenever there is a shared resource and multiple users.

Table 2.1
Details of the conceptual model of the queueing situation in Figure 2.6.

Element	Details
Entity	Queue, Server, Person
State Variables	Q = Number of Persons in Queue Q ∈ {0, 1, 2, ...} S = Status of Server S ∈ {Free, Busy}
Events	Arrival, Start_Service, End_Service (or Departure)
Activities	Generation, Waiting, Service, Delay

turn. Whenever the server becomes free, it will check the queue state variable, Q. If its value is greater than zero, the server will pick the next person from the queue and serve it. On the other hand, if the value of the queue state variable is zero, the server becomes idle because the queue is empty.

State variables are functions of time. Their evolution over time is referred to as a *sample path*. It can also be called a *realization* or *trajectory* of the state variable. Figure 2.8 shows one possible sample path of the state variable Q. Sample paths of DESs have a special shape which can be represented by a piecewise constant function. This function can also be referred to as a step function. In this kind of function, each piece represents a constant value that extends over a interval of time. The function changes its value when an event occurs. The time intervals are not uniform. They can be of different lengths. These observations are illustrated in Figure 2.8.

Four possible activities take place inside the single-server queueing system. They are shown in Figure 2.9. In the first activity, arrivals are generated into the system. This activity is bounded between two consecutive arrival events.

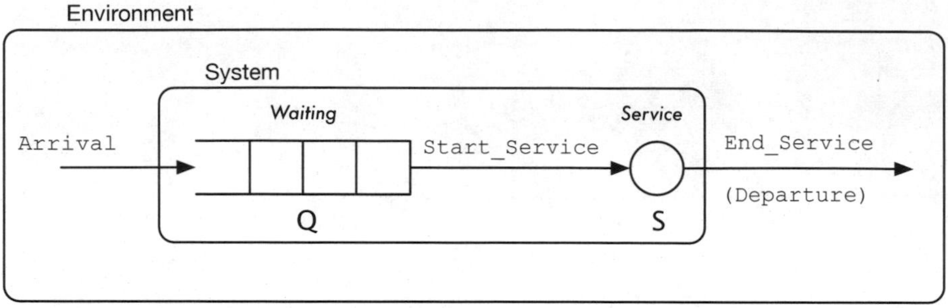

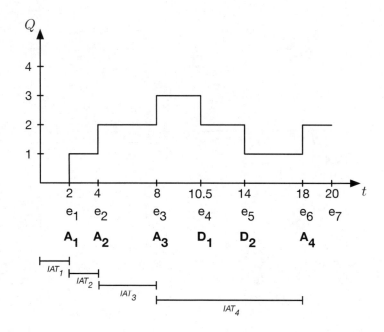

Figure 2.7
Conceptual model of the queueing situation in Figure 2.6.

Figure 2.8
A sample path of the state variable Q which represents the number of persons in the single-server queueing system. Note the difference in the time between every two consecutive arrival events.

The time between two such arrivals is random and it is referred to as the Inter-Arrival Time (IAT). This information is also shown in Figure 2.9.

The next activity involves waiting. This activity is initiated when an arriving person finds the server busy (i.e., $S = $ Busy). The waiting activity is terminatd when the server becomes free. Everytime the server becomes free, a Start_Service event is generated to indicate the start of service for the

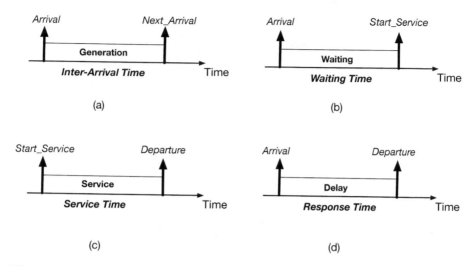

Figure 2.9
Four activities occur inside the single-server queueing system: (a) Generation, (b) Waiting, (c) Service, and (d) Delay. The length of each activity is a random variable of time.

next person in the queue. The difference between the time of the arrival of a person and his start of service is referred to as the Waiting Time (WT).

The third activity is about the time spent at the server. It is referred to as the Service Time (ST). This activity is initiated by a `Start_Service` event and terminated by an `End_Service` (or `Departure`) event, provided the two events are for the same person.

The length of the last activity is the total time a person spends in the system. It includes the waiting time and service time. This quantity represents the delay through the system or how long the system takes to respond to (i.e., fully serve) an arrival. This is the reason this quantity is also called the Response Time (RT). The events that start and terminate this activity are the `Arrival` and `Departure` events, respectively.

Modeling with Randomness

As you will learn in Chapter 4, the four activities in Figure 2.8 (IAT, WT, ST, and RT) can be modeled as random variables with probability distributions. Typically, the IAT and ST are modeled using exponential random variables. As for the other two activities (i.e., WT and RT), their probability distributions are not known in advance. They can be computed using simulation.

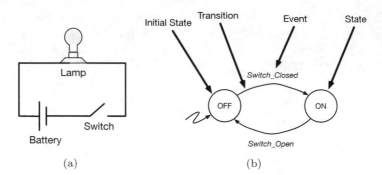

(a)　　　　　　　　　　　　　　(b)

Figure 2.10
A simple electrical circuit and its state diagram. Only the switch and lamp are modeled. Events are generated by the switch to change the state of the lamp.

2.4　STATE DIAGRAMS

A state diagram is a graph which is made up of circles representing states, arrows representing transitions between states, and a special (zigzag) arrow pointing to the initial state. The initial state is the state of the system at time zero. A transition between two states is caused by an event. The name of the event is typically placed above the arrow. The name of the state is placed inside the circle.

Figure 2.10 shows the state diagram for an electrical circuit consisting of a battery, switch, and lamp. Only the switch and lamp are modeled. The switch has two states: Open and Closed. Since we are interested in the lamp, the two states of the switch become the events that cause the lamp to change its state. The lamp has two states: On and Off. When the switch is closed, the lamp is on. When the switch is open, however, the lamp is off. As shown in Figure 2.10, this behavior can easily be captured by a state diagram.

Finally, it should be pointed out that each circle in a state diagram represents one possible realization of a state variable (or a group of state variables). That is, a state represents one possible value from the set of values the state variable can take. For example, in the case of the single-server queueing system in Figure 2.7, there are two state variables representing the states of the queue and server (see Table 2.1). The queue state variable is an integer variable that takes only positive values. The server state variable is a binary variable that takes two values only.

The state diagram for the queue shown in Figure 2.11(a) contains a circle for each positive integer number. Similarly, the state diagram for the server shown in Figure 2.11(b) contains only two circles. The initial states of the queue and server are 0 and Free, respectively. The queue state variable is incremented whenever an arrival event occurs. It is decremented whenever a

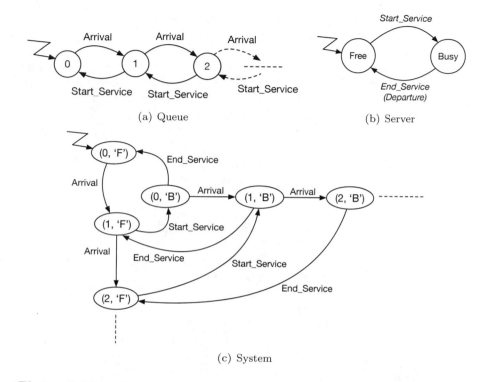

Figure 2.11
State diagrams of the state variables associated with the queue and server in the single-server queueing system in Figure 2.7. A portion of the state space of the system is shown in (c).

server event occurs. The server, on the other hand, becomes busy whenever a service starts. Then, at the end of the service, it becomes free again.

Figure 2.11(c) shows a portion of the state space of the system. The state diagram of the system combines the two state diagrams in Figures 2.11(a) and 2.11(b). In this new state diagram, each state is composed of two state variables (i.e., a state vector). Further, the state of the system is driven by three events: Arrival, Start_Service, and End_Service.

2.5 ACTUAL TIME VERSUS SIMULATED TIME

In simulation, we need to distinguish between two kinds of time. The first one is called the *actual time*. You are more familiar with this kind of time from your computer programming courses. It is the time a computer takes to fully execute a program. It is also referred to as the *runtime* of the program. The runtime is a function of the complexity of the conceptual model. For example, the execution time of your program will be large if you have many state variables, events, and activities in your model.

On the other hand, the *simulated time* is the time inside your conceptual model. It is not the time of the computer program that executes your model. Another name for this kind of time is *simulation time*. Simulation time does not pass at the same speed as actual time. That is, one second of simulated time is not necessarily equal to one second of actual time. In fact, they will be equal only in real-time simulation.

Because of this distinction between runtime and simulation time, you may simulate a phenomenon that lasts for a few years of actual time in one hour of simulation time. Similarly, you may simulate a phenomenon that lasts for a few seconds of simulated time in one hour of actual time.

2.6 SUMMARY

Five essential concepts used in building conceptual models have been covered in this chapter. Also, the famous single-server queueing system and its conceptual model have been introduced. The relationship between events and state variables has been shown using state diagrams. Finally, the difference between actual time and simulated time has been discussed. Clearly, this has been an important chapter, providing you with essential terms and tools in simulation.

2.7 EXERCISES

2.1 Consider a vending machine that accepts one, two, five, and ten dollar bills only. When a user inserts the money into a slot and pushes a button, the machine dispenses one bottle of water, which costs one dollar. The vending machine computes the change and releases it through another slot. If the vending machine is empty, a red light will go on.

   a. Identify the entities, state variables, events, and activities in this system, and

   b. Draw a state diagram which captures the behavior of the system.

2.2 Consider an Automated Teller Machine (ATM). A user of the ATM machine has to insert an ATM card and enter a Personal Identification Number (PIN). Both pieces of information must be validated before the main menu is displayed. If they are not valid, the card will be rejected. The ATM machine offers the following services:

   - Check balance,

   - Withdraw cash,

   - Deposit cash, and

 - Pay bills.

Answer the following questions:

a. Identify state variables, events, and activities in this system, and

b. Draw a state diagram that captures the behavior of the system.

## CHAPTER 3

Simulating Probabilities

"Probability does not exist."
−Bruno de Finetti

Probability is the science of uncertainty. It is used to study situations whose outcomes are unpredictable. In this chapter, first, the notion of probability is described. Then, it is shown how probability can be computed as a relative frequency using simulation. Finally, the sample mean is introduced as an estimate of probability.

3.1 RANDOM EXPERIMENTS AND EVENTS

Probability begins with the notion of a *random experiment*. A random experiment is an activity whose outcome is not known in advance. For example, as shown in Figure 3.1, the random experiment is about tossing a coin to observe what face turns up. This experiment has two possible outcomes: Head and Tail.

The set of all possible outcomes is referred to as the *sample space* of the random experiment. For the experiment of tossing a coin, the sample space, denoted by Ω, is the following:

$$\Omega = \{\text{Head}, \text{Tail}\}.$$

As another example, the sample space for the random experiment of throwing a die, which is shown in Figure 3.2, is as follows:

$$\Omega = \{1, 2, 3, 4, 5, 6\}.$$

The individual outcomes of a random experiment are not interesting by themselves. When one or more outcomes are combined, they form an event. In fact, any subset of the sample space is called an event. The following are some events that can be defined for the random experiment of throwing a die:

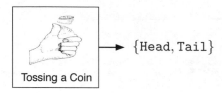

Figure 3.1
A random experiment of tossing a coin. There are two possible outcomes.

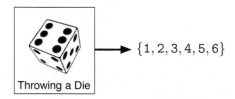

Figure 3.2
A random experiment of throwing a die. There are six possible outcomes.

- $E_1 = \{\text{One is observed}\} = \{1\}$,

- $E_2 = \{\text{A number less than four is observed}\} = \{1, 2, 3\}$, and

- $E_3 = \{\text{A number greater than or equal to five is observed}\} = \{5, 6\}$.

An event occurs whenever any of its outcomes are observed. For example, consider event E_1 above. This event occurs whenever we observe the outcome 1. Similarly, the event E_3 occurs if the outcome 5 or 6 is observed. Of course, both outcomes cannot be observed at the same time since the random experiment of throwing a die has only one outcome.

3.2   WHAT IS PROBABILITY?

A probability is a number between 0 and 1, inclusive. It is assigned to each possible outcome in the sample space of a random experiment. Probabilities must be assigned to outcomes in such a way that they add up to one. Once we have a valid assignment of probabilities to outcomes, the probability of an event is simply the sum of the probabilities of the individual outcomes making up the event.

For a discrete sample space, such as in the random experiments in Figures 3.1 and 3.2, a valid probability assignment can be achieved by assuming that all the outcomes are equiprobable. Therefore, the probability of each outcome can be computed as follows:

$$P(\omega_i) = \frac{1}{|\Omega|}, \quad \omega_i \in \Omega, \tag{3.1}$$

where $|\Omega|$ is the size of the sample space.

For a continuous sample space, we cannot assign probabilities to the individual elements in the sample space because they are not countable. Instead, we assign probabilities to regions of the sample space. For example, consider the one-dimensional region (or interval) $[a, b]$. The probability of any subregion $[j, k] \subseteq [a, b]$ can be defined as follows:

$$
\begin{aligned}
P([j, k]) &= \frac{\text{Length of } [j, k]}{\text{Length of } [a, b]} \\
&= \frac{|k - j|}{|b - a|}.
\end{aligned}
\tag{3.2}
$$

Hence, the probability of any one-dimensional region is proportional to its length. For a two-dimensional region, we use the area of the region to find its probability.

It should be pointed out that the above two rules for probability assignment cannot be used in simulation. As you will see in the next section, you will have to write a program that simulates the random experiment. In addition, your program should contain code for detecting the occurrence of the outcome or event of interest.[1] This is another difference between simulation and mathematical modeling. The probability is computed programmatically in simulation.

The next side note states four conditions that must be met in order to have a valid probability assignment. These conditions were developed by the Russian mathematician Andrey Kolmogorov in 1933. They are also called the axioms of probability.

---

[1] From now on, we are going to use the words "outcome" and "event" interchangeably.

Assigning Probabilities to Outcomes and Events

The following conditions must be satisfied in order to have a valid probability assignment.

1. For each outcome $\omega_i \in \Omega$, $P(\omega_i) \in [0, 1]$,

2. For all outcomes $\omega_i \in \Omega$, $\sum_i P(\omega_i) = 1$,

3. For each event $E_j \subseteq \Omega$, $P(E_j) = \sum_i P(\omega_i)$, where $\omega_i \in E_j$, and

4. For all possible disjoint events $E_j \subseteq \Omega$,

$$P(\{\bigcup_j E_j\}) = \sum_j P(E_j).$$

3.3 COMPUTING PROBABILITIES

In this section, you will learn how to compute probabilities programmatically. That is, you write a program which includes a model of your random experiment. Then, you run the program a sufficiently large number of times. The number of times the event of interest occurs is recorded. Then, this number is divided by the total number of times the random experiment is performed. The result is a probability that represents the *Relative Frequency* (RF) of the event of interest. The relative frequency of an event E_i is defined as follows.

$$P(E_i) = RF(E_i)$$
$$= \frac{\text{No. of times } E_i \text{ occurs}}{\text{No. of times the random experiment is performed}}. \tag{3.3}$$

As an example, consider again the random experiment of throwing a die. We want to approximate the probability of an outcome by means of a computer program. First, we need to learn how we can simulate this random experiment. Listing 3.1 shows how this random experiment can be simulated in Python. On line 1, the function *randint* is imported from the library *random*. That means the imported function becomes part of your program. The random experiment is represented by the function call *randint(1, 6)* on lines 3-5. Each one of these function calls return a random integer between 1 and 6, inclusive. The result of each function call is shown as a comment on the same line.

Now, after we know how to simulate the random experiment of throwing a die, we need to write a complete simulation program that contains the necessary code for checking for the occurrence of the event of interest and maintaining a counter of the number of times the event is observed. Also, the

Listing 3.1
Simulating the experiment of throwing a die. The output is shown as a comment on each line.

```python
from random import randint

print( randint(1,6) ) # outcome = 1
print( randint(1,6) ) # outcome = 3
print( randint(1,6) ) # outcome = 5
```

Listing 3.2
Approximating the probability of an outcome in the experiment of throwing a die.

```python
from random import randint

n = 1000000     # No. of times experiment is performed
ne = 0          # Count the occurrences of event

for i in range(n):
    outcome = randint(1, 6)
    if(outcome == 3):        # Check for event of interest
        ne += 1              # ne = ne + 1

print("Prob = ", round(ne / n, 4))  # = 0.1667
```

program should compute the probability of the event using Eqn. (3.3). This program is shown as Listing 3.2.

The program has four parts. In the first part (line 1), the function *randint* is included into the program. This function is used to simulate the random experiment of throwing a die. Next, in the second part (lines 3-4), two variables are defined. The first one is a parameter whose value is selected by you before you run the program. It represents the number of times the random experiment is performed. The second variable is an event counter used while the program is running to keep track of the number of times the event of interest is observed. The third part of the program contains the *simulation loop* (line 6). Inside this loop, the random experiment is performed and its outcome is recorded (line 7). Then, a condition is used to check if the generated outcome is equal to the event of interest (line 8). If it is indeed equal to the event of interest, the event counter is incremented by one. The experiment is repeated n number of times. Finally, in the last part of the program, the probability of the event of interest is computed as a relative frequency (line 11). The function *round* is used to round the probability to four digits after the decimal point.

The function *round* is not explicitly included into the program. This is because it is a built-in function. Python has a group of functions referred to as the built-in functions which are always available. Some of these functions are *min*, *max*, and *len*. You are encouraged to check the Python documentation for more information.[2]

3.4 PROBABILITY AS A SAMPLE MEAN

Let us say that you want to *estimate* the probability of seeing a head in the random experiment of tossing a coin (see Figure 3.1). The *theoretical* (also called *true*) value of this probability is $\frac{1}{2}$. We are going to use the relative frequency as our *estimator* (see Eqn. (3.3)) and the event of interest is *Head*. The first part of Listing 3.3 shows how this probability can be estimated using simulation. In each iteration of the simulation loop (8-13), if a head is observed, a one is added to the list *observed*. At the end of the simulation loop, the *mean* (also called the *average*) of this list is computed using the *mean* function from the *statistics* library. The mean obtained in this way is called the *sample mean*. This is because if this part of the program is run again, a different mean will result. Note that the outcomes *Head* and *Tail* have to be encoded using integer numbers; otherwise, the function *mean* cannot be used. Also, note that in each run of the program, the list *observed* will contain one sample.

The above description is illustrated in Figure 3.3. In this case, the experiment is performed 10 times only. The axis at the top keeps track of the iteration number. The figure shows three samples where each is of size 10. The three samples result in three different means. The sample mean can mathe-

---

[2]https://docs.python.org/3/library/functions.html.

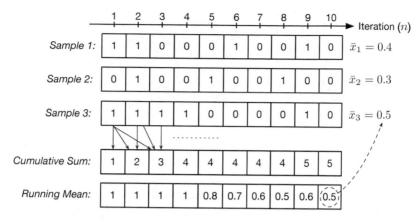

Figure 3.3
Three different samples for the random experiment of tossing a coin 10 times.
The running mean is computed for the third sample using cumulative sums.
The last value at position 10 of the list of running means is equal to the sample
mean. The sample mean is the probability of seeing a head.

matically be defined as follows:

$$\bar{x} = \frac{\sum_i^n x_i}{n}, \tag{3.4}$$

where n is the size of the sample. Each entry in the sample represents the
outcome of a single iteration of the random experiment.

Clearly, each sample has to be long enough (i.e., large n) in order to obtain
a good estimate of the probability of seeing a head. This fact is shown in Figure
3.4. For 100 iterations, the probability curve is still very far away from the
true value. However, for 1000 iterations, the probability curve converges to
the true value.

In order to show the behavior of the sample mean over the different iterations of the random experiment (like in Figures 3.4(a) and (b)), we record the
running mean at the end of each iteration. Figure 3.3 shows how the running
mean is calculated by first calculating the cumulative sums, which are then
divided by the number of samples used in each sum. The running mean can
be defined as follows:

$$\bar{x}_j = \frac{\sum_i^j x_i}{j}, \tag{3.5}$$

where j is the iteration number, x_i is the integer value corresponding to the
outcome of the random experiment at the end of the jth iteration ($i \leq j$), and
$\sum_i^j x_i$ is the jth cumulative sum. The second part of Listing 3.3 shows how
the running mean is computed. Then, in the third part, it is shown how the
plots in Figure 3.4 are produced.

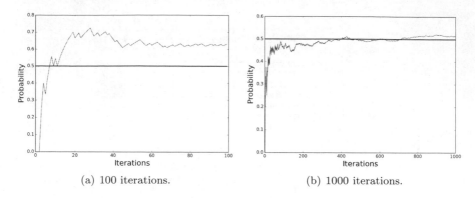

(a) 100 iterations.   (b) 1000 iterations.

Figure 3.4
Running mean for the random experiment of tossing a coin. The mean eventually converges to the true value as more samples are generated.

Population Versus Sample

In statistics, the term "population" refers to a set of all possible realizations of your random experiment. If you are going to toss a coin only once, there are two possibilities: *Head* or *Tail*. However, if you are going to perform the same experiment 10 times (see Figure 3.3), then the size of the population is $2^{10} = 1024$. This number represents the number of all possible realizations of the experiment of tossing a coin 10 times. Three samples from this population are shown in Figure 3.3.

Note that as the number of iteration increases, the sample mean converges to a specific value, which is the true mean. Before this convergence happens, the sample mean will fluctuate. A large number of samples will be needed before the mean stabilizes and hits the true mean. The true mean is referred to also as the *population* mean. The the difference between the population and sample means is explained in the above side note. The use of the sample mean as a probability is supported by the law of large numbers stated in the next side note.

The Law of Large Numbers

Let x_1, x_2, ..., x_n be a set of samples (realizations / outcomes of a random experiment), then with probability 1

$$\lim_{n \to \infty} \frac{x_1 + x_2 + ... + x_n}{n}$$

converges to the population mean. That is, the running mean will eventually converge to the population mean.

Listing 3.3
Simulation program for studying the running mean of the random experiment of tossing a coin. This program is also used to generate Figure 3.4.

```python
### Part 1: Performing the simulation experiment
from random import choice
from statistics import mean

n = 1000
observed = []

for i in range(n):
    outcome = choice(['Head', 'Tail'])
    if outcome == 'Head':
        observed.append(1)
    else:
        observed.append(0)

print("Prob = ", round(mean(observed), 2))

### Part 2: Computing the moving average
from numpy import cumsum

cum_observed = cumsum(observed)

moving_avg = []
```

```
23  for i in range(len(cum_observed)):
24    moving_avg.append( cum_observed[i] / (i+1) )
25
26  ### Part 3: Making the plot
27  from matplotlib.pyplot import *
28  from numpy import arange
29
30  x = arange(0, len(moving_avg), 1) # x-axis
31  p = [0.5 for i in range(len(moving_avg))] # Line
32
33  xlabel('Iterations', size=20)
34  ylabel('Probability', size=20)
35
36  plot(x, moving_avg)
37  plot(x, p, linewidth=2, color='black')
38
39  show()
```

3.5 SUMMARY

In this chapter, you have learned how to build simulation models for simple random experiments. You have also learned how to write a complete simulation program that includes your simulation model in the Python programming language. Further, you have been exposed to the essence of simulation, which is estimation. As has been shown in Figure 3.4, the length of a simulation run (i.e., the value of n) plays a significant role in the accuracy of the estimator.

3.6 EXERCISES

3.1 Consider the random experiment of throwing two dice. What is the probability of the event that three spots or less are observed? Show how you can compute the probability of this event both mathematically and programmatically.

3.2 Write a Python program to simulate the random experiment of tossing a fair coin five times. Your goal is to estimate the probability of seeing four heads in five tosses.

a. What is the size of the population?

b. List all the possible realizations of the experiment which contain four heads.

c. Use the information obtained in (b) to mathematically compute the probability of the event of interest.

d. Write a Python program to estimate the same probability. How many times do you need to repeat the experiment to get a good estimate of the probability (i.e., what should be the size of n)?

3.3 Consider a class of 20 students. Show how you can programmatically compute the probability that at least two students have the same birthday. (*Answer* $= 0.41$).

3.4 A box contains two white balls, two red balls, and a black ball. Two balls are randomly chosen from the box. What is the probability of the event that the second ball is white *given* that the first ball chosen is white? Use simulation to approximate this probability. (*Answer* $= 0.25$)

3.5 You are playing a game with a friend. Your friend flips a coin 7 times and you flip a coin 8 times. The person who gets the largest number of tails wins. Your friend also wins if you both get the same number of tails. Write a simulation program that estimates the probability that you win this game. (*Answer* $= 0.5$)

3.6 Write a Python program to simulate the following game. Three people take turns at throwing a die. The number of dots on each face of the die represents a number of points. The first person who accumulates 100 points win the game.

3.7 Consider a password generator which generates a string of size N. The generated string should contain letters and numbers. One of the letter has to be in uppercase. The following are sample passwords produced by this password generator: "90MJK" and "OLM15ndg". Write a Python program for this password generator.

## CHAPTER 4

Simulating Random Variables and Stochastic Processes

"Mathematics is written for mathematicians."
—Nicolaus Copernicus

A dynamic system can be viewed as either a sequence of random variables or a stochastic process. In the former, the system evolves in discrete steps and its state changes at discrete instants of time. In the latter, however, the system evolves continuously over time and the times at which events occur cannot be predicted. In both cases, activities inside a simulation model can be thought of as random variables. Random variables are mathematical abstractions which are defined on the sample space of a random experiment. Another important mathematical abstraction is stochastic (or random) processes, which extend random variables by introducing time as another source of variability. A simulation model evolves as a stochastic process. This chapter discusses random variables and stochastic processes in the context of discrete-event simulation.

4.1 WHAT ARE RANDOM VARIABLES?

A random variable is a function which associates a probability with each possible outcome of a random experiment. The domain of a random variable is a set of numerical values which correspond to events defined on the sample space of the random experiment. For example, in the experiment of throwing two fair dice, one possible event is that the two dice show up $\{1, 1\}$. This event is represented as $\{X = 2\}$. The range of a random variable, on the other hand, is a probability (e.g., $P[X = 2] = \frac{1}{36}$). Figure 4.1 illustrates the

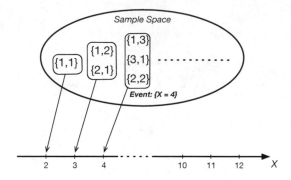

Figure 4.1
Sample space for the random experiment of throwing two dice. The outcome
of the experiment is a random variable $X \in \{2, 3, ..., 12\}$.

previous example. Events are encoded as integer numbers. In this way, we
abstract away the details of the sample space.

For convenience, a random variable is denoted by a capital letter (e.g., X).
Its value, however, is denoted by a small letter (e.g., x). A random variable
is either *discrete* or *continuous*. A discrete random variable has a finite set
of values. For example, the number of customers in a waiting line is discrete.
By contrast, a continuous random variable has an infinite set of values, which
cannot be mapped onto the integers. The time a customer spends in a waiting
line is an example of a continuous random variable.

A simple rule to decide whether you should use a discrete or continuous
random variable is as follows. If the quantity you want to include in your model
can be counted like the number of cars and number of servers, then you should
use a discrete random variable. On the other hand, if the quantity cannot be
counted (i.e., you need a measuring device to tell the exact value), then you
should use a continuous random variable. For example, to tell the temperature
of a patient, you need a thermometer. So, temperature is a continuous quantity
and should be represented by a continuous random variable.

Next, probability functions are discussed. Probability functions are used to
characterize random variables. In Chapter 10, they will be used for the purpose
of simulating random variables. For now, however, you are going to use the
pre-built functions from the *random*[1] and *scipy.stats*[2] libraries whenever you
want to simulate a random variable.

4.1.1 Probability Mass Functions

For a discrete random variable, the probability function is referred to as the
Probability Mass Function (PMF). For each possible value of the random

---

[1] https://docs.python.org/library/random.html.
[2] http://docs.scipy.org/doc/scipy/reference/stats.html.

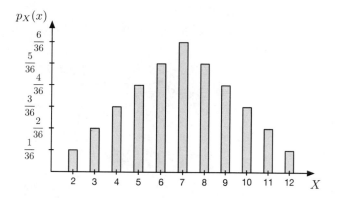

Figure 4.2
The PMF of a discrete random variable representing the outcome of the random experiment of throwing two dice.

variable, the PMF associates a probability with it. The PMF is denoted by $p_X(x)$ and it is defined as follows:

$$p_X(x) = P[X = x]. \tag{4.1}$$

The PMF satisfies the following two properties:

1. $p_X(x) \geq 0$, and

2. $\sum_x p_X(x) = 1$.

Hence, because of the above two properties, the PMF is a probability function. Figure 4.2 shows the PMF of a random variable representing the outcome of the random experiment of throwing two dice. The length of each bar represents a probability. There are gaps between the bars since they correspond to discrete values.

4.1.2 Cumulative Distribution Functions

The Cumulative Distribution Function (CDF) of a discrete random variable X, denoted by $F_X(x)$, is mathematically defined as follows for $-\infty < X < +\infty$:

$$F_X(x) = P(X \leq x)$$
$$= \sum_{i \leq x} p_X(i). \tag{4.2}$$

Basically, the CDF gives the probability that the value of the random variable X is less than or equal to x. Thus, it is a monotonically non-decreasing function of X. That is, as the value of X increases, $F_X(x)$ increases or stays the same. Figure 4.3 shows the CDF of the random variable representing the experiment

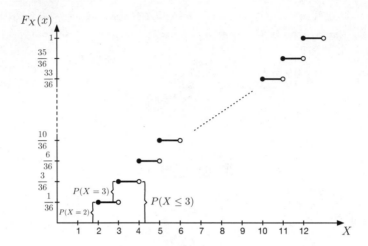

Figure 4.3
The cumulative distribution function of a discrete random variable representing the outcome of the random experiment of throwing two dice.

of throwing two dice. As an example, the probability of the event $\{X \leq 5\}$ is computed as follows:

$$
\begin{aligned}
P[X \leq 5] &= F_X(5) \\
&= \sum_{i \leq 5} p_X(i) \\
&= p_X(2) + p_X(3) + p_X(4) + p_X(5) \quad\quad (4.3)\\
&= \frac{1}{36} + \frac{2}{36} + \frac{3}{36} + \frac{4}{36} \\
&= \frac{10}{36}.
\end{aligned}
$$

As Figure 4.3 shows, the CDF of a discrete random variable is not continuous. It is rather a staircase function. This kind of function is drawn as shown in Figure 4.3. Each line segment corresponds to a probability and a range of values for X. Since X is discrete, each segment corresponds to one value only. Also, each segment has two endpoints: Closed and Open. The closed endpoint which is represented by a black dot indicates that the corresponding value on the x-axis is part of the line segment. On the other hand, the open endpoint which is represented by an open circle indicates that the corresponding value on the x-axis is not part of the line segment. For example, consider the

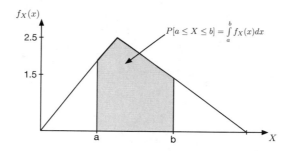

Figure 4.4
Probability density function of a continuous random variable.

probability of the event $\{X = 2\}$:

$$
\begin{aligned}
P[\{X = 2\}] &= P[\, X \in [2,3)\,] \\
&= P[X = 2] \\
&= p_X(2) \\
&= \frac{1}{36}.
\end{aligned}
\tag{4.4}
$$

4.1.3 Probability Density Functions

A continuous random variable has a special function referred to as the Probability Density Function (PDF). The PDF is a non-negative, continuous function denoted by $f_X(x)$. It is not a probability function because it can be greater than one. Therefore, the following is wrong:

$$
\begin{aligned}
f_X(x) &= P[\, X = x\,] \\
&= \int_x^x f_X(x)dx \\
&= 0.
\end{aligned}
\tag{4.5}
$$

Instead, the CDF is used whenever a probability of a continuous random variable is required.

Consider the PDF in Figure 4.4. When integrated, the PDF gives the probability of X belonging to a finite interval. This can be mathematically expressed as follows:

$$
\begin{aligned}
P[\{a \le X \le b\}] &= P[\, X \in [a,b]\,] \\
&= \int_a^b f_X(x)dx \\
&= F_X(b) - F_X(a).
\end{aligned}
\tag{4.6}
$$

As will be shown in Chapter 9, the probability in Eqn. (4.6) can be approximately computed using the Monte Carlo method.

The CDF for a specific value is computed as follows:

$$P[\{X \leq i\}] = F_X(i)$$

$$= \int_{-\infty}^{i} f_X(x)dx. \tag{4.7}$$

The following are the relationships between the CDF and its PDF:

$$f_X(x) = \frac{d}{dx}F_X(x) \tag{4.8}$$

$$F_X(x) = \int_{-\infty}^{+\infty} f_X(x)dx. \tag{4.9}$$

The reader is encouraged to refer to his college textbooks on calculus to remind himself of the basics of differentiation and integration.

4.1.4 Histograms

A histogram is a graph that shows the distribution of data in a data set. By distribution, we mean the frequency (or relative frequency) of each possible value in the data set. A histogram can be used to approximate the PDF of a continuous random variable. It can also be used to construct the PMF of a discrete random variable.

The range of values in a data set represents an interval. This interval can be divided into subintervals. In a histogram, each subinterval is represented by a bin on the x-axis. On each bin, a bar is drawn. The length of the bar is relative to the number of samples (i.e., data values) in the corresponding bin. The area of the bar is thus the product of its length and the width of the bin. This quantity is equal to the probability that a sample falls in the subinterval represented by the bin. Figure 4.5 illustrates the common elements of a histogram.

Listing 4.1
Python program for generating the histogram from an exponential data set (see Figure 4.6).

```
1  from random import expovariate
2  from matplotlib.pyplot import hist, xlabel, ylabel, title,
       show, savefig
3
```

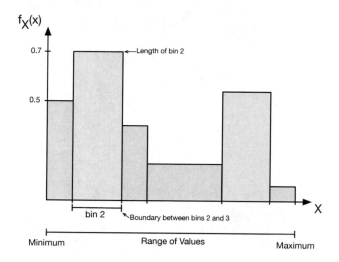

Figure 4.5
Elements of a histogram. Bins can be of different widths. Length of a bar
could represent frequency or relative frequency.

```
4   # Generate the data set
5   N = 10000
6   data = [expovariate(1.5) for i in range(N)]
7
8   # Decide number of bins
9   num_bins = 50
10
11  # Construct the histogram of the data
12  # To generate PDF, use normed=True
13  n, bins, patches = hist(data, num_bins, normed=True,
        facecolor='black', alpha=0.6)
14
15
16  xlabel('$X$', size=18)
17  ylabel('$f_X(x)$', size=18)
18  title('Histogram of exponential data: $\mu$ = 1.5', size=15)
19
20  # Show the figure or  save it
```

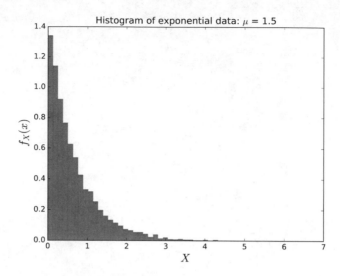

Figure 4.6
Histogram for an exponential data set. This figure is generated using Listing 4.1.

```
21  #show()
22  savefig('hist_expov.pdf', format='pdf', bbox_inches='tight')
```

Listing 4.1 shows how a histogram can be constructed and plotted in Python. First, a random data set is generated on line 6. Then, the number of bins in the histogram is decided on line 9. After that, the function *hist* is called to construct and plot the histogram (see line 13). The argument *normed* is set to `True` to generate relative frequency. This is important if you want to approximate the PDF of a random variable. The rest of the program is clear from the previous examples. Figure 4.6 shows the histogram generated using this example.

4.2 SOME USEFUL RANDOM VARIABLES

4.2.1 Bernoulli

A Bernoulli random variable is a *discrete* random variable that can be used for modeling random experiments with two outcomes only. The outcome of interest is typically referred to as a success and it is represented by the integer number 1. As an example, the outcome of the random experiment of tossing a coin can be modeled as a Bernoulli random variable. In this experiment, the

---

Indicator Functions

An indicator function is denoted by the symbol $\mathbb{1}$ with a subscript $\mathbb{E}$ describing the event of interest. If the event is observed, the function returns 1; otherwise, it returns 0.

$$\mathbb{1}_{\mathbb{E}} = \begin{cases} 1, & \text{if } \mathbb{E} \text{ } occurs, \\ \\ 0, & otherwise. \end{cases}$$

Indicator functions are used in transforming probability functions defined over two lines to one which is defined only on one line. The following example shows how the PMF of the Bernoulli random variable in Eqn. (4.10) can be written on one line.

$$p_X(x) = p \cdot \mathbb{1}_{\{x=1\}}.$$

---

event of interest is Head and it is observed with a success probability of p. The following is the PMF of the Bernoulli random variable:

$$p_X(x) = \begin{cases} p, & \text{if } x = 1, \\ \\ 1 - p, & \text{if } x = 0. \end{cases} \qquad (4.10)$$

The following are the mean and variance, respectively:

$$\mu = p \qquad (4.11)$$

$$\sigma^2 = p(1 - p). \qquad (4.12)$$

The above side note explains how the PMF of the Bernoulli random variable (i.e., Eqn (4.10)) can be written on one line using an indicator function.

4.2.2 Binomial

The binomial random variable is an extension of the Bernoulli random variable, where the number of trials n is another parameter of the new random experiment. Basically, the Bernoulli experiment (or trial) is repeated n times. Then, the number of successes X in n trials is given by the following PMF:

$$p_X(x) = \binom{n}{x} p^x (1 - p)^{n-x}, \qquad (4.13)$$

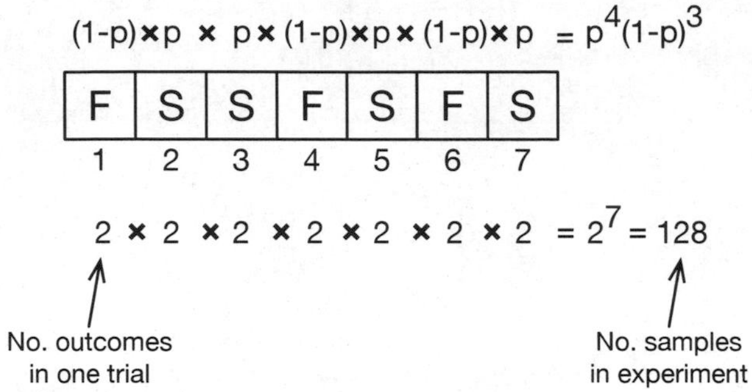

$$(1\text{-}p)\times p \;\times\; p\times (1\text{-}p)\times p\times (1\text{-}p)\times p \;=\; p^4(1\text{-}p)^3$$

F	S	S	F	S	F	S
1	2	3	4	5	6	7

$$2 \times 2 \times 2 \times 2 \times 2 \times 2 \times 2 \;=\; 2^7 = 128$$

No. outcomes
in one trial

No. samples
in experiment

Figure 4.7
The situation of observing four successes in a sequence of seven Bernoulli trials
can be modeled as a binomial random variable.

where p is the probability of success in a single trial. The following are the
mean and variance, respectively:

$$\mu = np \tag{4.14}$$

$$\sigma^2 = np(1 - p). \tag{4.15}$$

Figure 4.7 shows one sample in the sequential random experiment of seven
Bernoulli trials. This figure also illustrates how the probability associated with
each sample is calculated. Further, notice how the total number of samples
is calculated. Clearly, the number of samples exponentially increases with n.
This is why enumerating (i.e., listing) is a hard problem.[3] Instead, techniques
such as the Monte Carlo method in Chapter 9 are used to generate samples
intelligently.

4.2.3 Geometric

The random experiment of repeating a Bernoulli trial until the first success
is observed is modeled by a geometric random variable. This random variable
can also be defined as the number of failures until the first success occurs. The
PMF for a geometric random variable is the following:

$$p_X(x) = p(1 - p)^x, \tag{4.16}$$

where p is the probability of success in a single Bernoulli trial and $x \in
\{0, 1, 2, ...\}$. The following are the mean and variance, respectively:

$$\mu = \frac{1 - p}{p} \tag{4.17}$$

---

[3]In computer science, a problem is referred to as NP-hard if the runtime of any algorithm
that solves it is exponential (i.e., $\mathcal{O}(2^n)$).

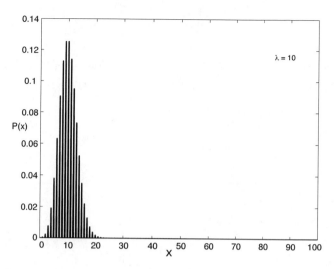

Figure 4.8
The PMF of the Poisson random variable for $\lambda = 10$. Notice that $P(x)$ approaches zero as x increases.

$$\sigma^2 = \frac{1-p}{p^2}. \tag{4.18}$$

4.2.4 Poisson

A Poisson random variable X is a *discrete* random variable which has the following probability mass function.

$$P(X = x) = \frac{\lambda^x \cdot e^{-\lambda}}{x!}, \tag{4.19}$$

where $P(X = x)$ is the probability of x events occurring in an interval of preset length, λ is the expected number of events (i.e., mean) occurring in the same interval, $x \in \{0, 1, 2, ...\}$, and e is a constant equal to 2.72. Figure 4.8 shows the PMF of the Poisson random variable for $\lambda = 10$.

The Poisson random variable can be used to model the number of frames[4] that arrive at the input of a communication system. The length of the observation interval must be specified when giving λ (e.g., five frames per 10 milliseconds which is equal to 0.5 frame per one millisecond). The next side note elaborates more.

Using the Poisson Random Variable for Modeling

The Poisson random variable is used to represent the number of events that occur in an interval of time of fixed duration throughout the random experiment. For instance, if frames arrive at a switch at an average rate of 15 per 50 minutes, then the probability of x arrivals in 50 minutes is calculated as follows:

$$P(X = x) = \frac{15^x \cdot e^{-15}}{x!}.$$

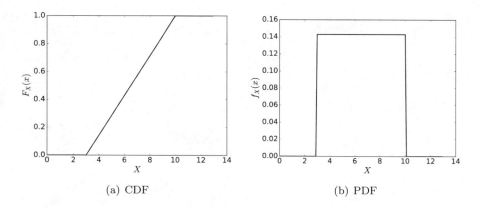

(a) CDF   (b) PDF

Figure 4.9
Probability distribution functions for the uniform random variable where $a = 3$ and $b = 10$.

4.2.5   Uniform

A uniform random variable X is a *continuous* random variable that has the following cumulative distribution function.

$$F(x) = \frac{x - a}{b - a}, \tag{4.20}$$

where $x \in [a, b]$. The probability density function is

$$f_X(x) = \begin{cases} \frac{1}{b-a}, & \text{for} \quad x \in [a, b], \\ \\ 0, & \text{otherwise.} \end{cases} \tag{4.21}$$

---

[4]Do you know that packets cannot travel through the wire? Actually, frames are the data units that travel through wires and they carry packets. This is why we use frames as our data unit.

Figures 4.9(a) and 4.9(b) shows the CDF and PDF of uniform random variable with $a = 3$ and $b = 10$. Listing 4.2 is the program used to generate these two figures.

Listing 4.2
Python program for plotting the CDF and PDF of a uniform random variable (see Figures 4.9(a) and 4.9(b)).

```python
from numpy import *
from matplotlib.pyplot import *


# Parameters
a = 3
b = 10


# Plotting the PDF
def pdf(x):
    if x >= a and x <= b:
        return 1 / (b - a)
    else:
        return 0

X = arange(0, b+3, 0.1)
Y = []


for x in X:
    Y.append(pdf(x))


matplotlib.rc('xtick', labelsize=18)
matplotlib.rc('ytick', labelsize=18)
plot(X, Y, Linewidth=2, color='black')
xlabel('$X$', size=22)
ylabel('$f_X(x)$', size=22)
#show()
```

```
27  savefig('uniform_pdf.pdf', format='pdf', bbox_inches='tight'
        )

28

29  # Clear the current figure
30  clf()

31

32  # Plotting the CDF
33  def cdf(x):
34      if x < a:
35          return 0
36      elif x >= a and x < b:
37          return (x - a) / (b - a)
38      elif x >= b:
39          return 1

40

41  X = arange(0, b+3, 0.1)
42  Y = []

43

44  for x in X:
45      Y.append(cdf(x))

46

47  matplotlib.rc('xtick', labelsize=18)
48  matplotlib.rc('ytick', labelsize=18)
49  plot(X, Y, Linewidth=2, color='black')
50  xlabel('$X$', size=22)
51  ylabel('$F_X(x)$', size=22)
52  #show()
53  savefig('uniform_cdf.pdf', format='pdf', bbox_inches='tight'
        )
```

The mean and variance of the uniform random variable are the following, respectively:

$$\mu = \frac{1}{2}(a + b) \tag{4.22}$$

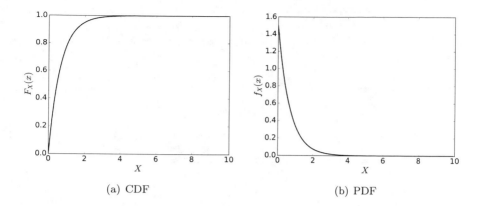

(a) CDF

(b) PDF

Figure 4.10
Probability distribution functions of the exponential random variable where $\mu = 1.5$.

$$\sigma^2 = \frac{1}{12}(b - a)^2. \tag{4.23}$$

This random variable is typically used to model equally likely events, such as the random selection of one item from a list of candidate items. The events can be modeled as equally likely because the PDF is constant for all the possible values of the random variable. This is why it is called a uniform random variable.

4.2.6 Exponential

An exponential random variable X is a *continuous* random variable which has the following cumulative distribution function.

$$F_X(x) = 1 - e^{-\mu x}, \tag{4.24}$$

where μ is the rate parameter and $x \in [0, \infty)$. On the other hand, the probability density function is given by the following expression:

$$f_X(x) = \mu e^{-\mu x}. \tag{4.25}$$

Figures 4.10(a) and 4.10(b) show the shapes of these two functions. Notice the initial value of the PDF. It is greater than one. This is normal since the PDF is not a probability function.

The exponential random variable can be used to model the time between the occurrences of two consecutive events. For example, it is used to model the time between two consecutive arrivals or departures in the single-server queueing system. The next side note explains the relationship between the Poisson and exponential random variables.

The Relationship between the Poisson and Exponential Random Variables

The time between two consecutive events occurring in an interval of fixed duration is modeled as an exponential random variable and it has the following PDF:

$$f_X(x) = \lambda e^{-\lambda x},$$

where λ is the average number of events occurring during the fixed interval of observation.

4.2.7 Erlang

The Erlang random variable is continuous. It can be expressed as a sum of exponential random variables. This property will be used in Section 10.4 to generate samples from the Erlang distribution. The Erlang random variable has two parameters:

1. Scale or rate (θ), and

2. Shape (k).

k is an integer and it represents the number of independent exponential random variables that are summed up to form the Erlang random variable. Hence, the Erlang distribution with k equal to 1 simplifies to the exponential distribution.

The following are the probability density and cumulative distribution functions of the Erlang random variable X:

$$f(x) = \frac{x^{k-1}\theta^k e^{-\theta x}}{(k-1)!}, \qquad x \geq 0 \tag{4.26}$$

$$F(x) = 1 - e^{-\theta x} \sum_{j=0}^{k-1} \frac{(\theta x)^j}{j!}, \qquad x \geq 0. \tag{4.27}$$

4.2.8 Normal

A normal (or Gaussian) random variable is a continuous random variable that has the following probability density function.

$$f(x) = \frac{1}{\sigma\sqrt{2\pi}} \cdot e^{-\frac{(x-\mu)^2}{2\sigma^2}} \tag{4.28}$$

where μ is the mean, σ is the standard deviation, and $x \in (-\infty, \infty)$. Figure 4.11 shows the shape of the PDF of the normal random variable. If $\mu = 0$ and $\sigma = 1$, the resulting PDF is referred to as the standard normal distribution and the resulting random variable is called the standard normal random variable.

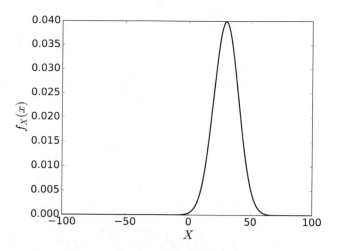

Figure 4.11
The PDF of the normal random variable with $\mu = 30$ and $\sigma = 10$.

4.2.9   Triangular

A triangular random variable has three parameters: a, b, and c. The last parameter is referred to as the *mode*. At this point, the PDF has the highest density. The following is the CDF:

$$F_X(x) = \begin{cases} 0, & \text{if } x \leq a, \\[2mm] \frac{(x-a)^2}{(b-a)(c-a)}, & \text{if } a < x \leq c, \\[2mm] 1 - \frac{(b-x)^2}{(b-a)(b-c)}, & \text{if } c < x < b, \\[2mm] 1, & \text{if } x \geq b. \end{cases} \tag{4.29}$$

The PDF is defined as follows:

$$f_X(x) = \begin{cases} 0, & \text{if } x < a, \\[2mm] \frac{2(x-a)}{(b-a)(c-a)}, & \text{if } a \leq x < c, \\[2mm] \frac{2}{b-a}, & \text{if } x = c, \\[1mm] \frac{2(b-x)}{(b-a)(b-c)}, & \text{if } c < x \leq b, \\[2mm] 0, & \text{if } x > b. \end{cases} \tag{4.30}$$

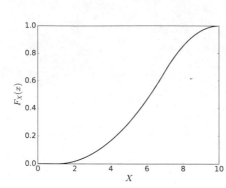

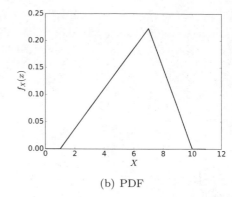

(a) CDF   (b) PDF

Figure 4.12
Probability distribution functions of the triangular random variable with $a = 1$, $b = 10$, and $c = 7$.

Figures 4.12(a) and 4.12(b) show the shapes of the CDF and PDF, respectively. The expected value of a triangular random variable X is

$$\mu = \frac{a+b+c}{3} \tag{4.31}$$

and the variance is

$$\sigma^2 = \frac{a^2 + b^2 + c^2 - ab - ac - bc}{18}. \tag{4.32}$$

4.3 STOCHASTIC PROCESSES

A random variable cannot be used to describe the behavior of a dynamic system since it does not involve time. How do you think time should be handled? Enter the world of stochastic processes. Figure 4.13 shows that a stochastic process is a collection of random variables whose indexes are discrete points in time. At every instant of time, the state of the process is random. And, since time is fixed, we can think of the state of the process as a random variable at that specific instant. At time t_i, the state of the process is determined by performing a random experiment whose outcome is from the set Ω. At time t_j, the same experiment is performed again to determine the next state of the process. That is the essence of stochastic processes.

A stochastic process can evolve in many different directions. Each direction is referred to as a *realization, trajectory,* or *sample path* of the process. Two sample paths are shown in Figure 4.13. They are $g_1(t)$ and $g_2(t)$. They are different from the time functions f_i. Multiple time functions are combined to construct one sample path.

A stochastic process can have two means: vertical and horizontal. The

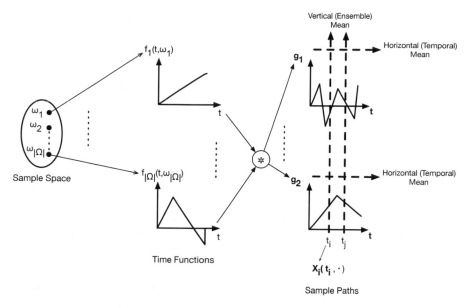

Figure 4.13
A stochastic process maps each outcome in the sample space to a time function. Time functions are combined (convoluted) to produce two sample paths: g_1 and g_2. Two kinds of means can be defined for a stochastic process.

What Do You Get When You Fix One Source of Variability?

In a stochastic process, when you fix time, you get a random variable. For example, in Figure 4.13, at time t_i, a random variable $X_i(t_i, \cdot)$ can be defined. Similarly, if you fix the outcome, you get a sample path, e.g., $f_1(\cdot, \omega_1)$.

---

vertical mean is called the *ensemble* mean. It is calculated over all the possible sample paths. The horizontal mean, however, is calculated using one sample path. This is why it is referred to as a *time* average. Fortunately, as you will learn in the next section, the horizontal mean can be used as an approximation of the vertical mean.

The Bernoulli random process is illustrated in Figure 4.14. This process is composed of two time functions, which are both constant (see Figure 4.14(c)). Figure 4.14(c) shows the result of running the fundamental Bernoulli random experiment in each trial (i.e., time slot). The function $f(t)$ in Figure 4.14(d) does not represent the real behavior of the Bernoulli random process. However, it is used to construct this behavior in Figure 4.14(e). The Bernoulli random process is a counting process. It counts the number of ones observed so far.

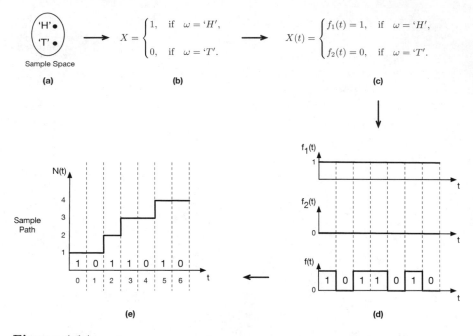

Figure 4.14

The Bernoulli random process: (a) sample space, (b) random variable, (c) time functions, (d) result of running the random experiment in each slot, and (e) final sample path.

4.4 DYNAMIC SYSTEM EVOLUTION

Since a dynamic system can be viewed as a stochastic process, it can have many sample paths. Figure 4.15 shows an example of the state evolution of a system. In this figure, each circle represents a possible state of the system. The state of the system can be thought of as a vector of state variables. For example, as we will see in the next chapter, the single-server queueing system has two state variables: (1) number of customers present in the queue and (2) Status of the server (either busy or idle). These two variables represent all the information we need to study this system. Finally, it should be pointed out that the diagram in Figure 4.15 is referred to as the state space because it contains all the possible states of the system.

When a system is simulated, each simulation run represents an evolution of the system along one path in the state space. That is, a sample path $g_i(t)$, where i is the index of the simulation run. Figure 4.16 shows an example wherein six events have been simulated in the order shown in the figure. This sample path represents one possible behavior of the system.

As you will see in Chapter 11, the data generated and collected along one trajectory in the state space is used in the estimation of the performance measure of interest. Of course, an estimate will be more accurate if more

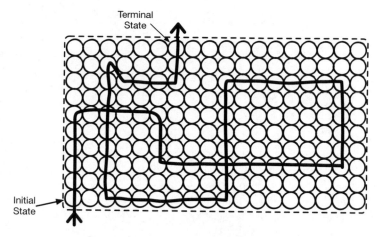

Figure 4.15
A sample path through the state space of a dynamic system. Entry and exit points are random. Data is generated along this sample path and a time average is computed as an estimate of the performance metric of interest.

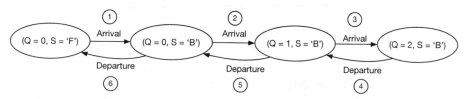

Figure 4.16
A sample path through the state space of the single-server queueing system. The initial state does not have to be (0, 'F').

simulation runs are performed and their average is used instead. Hopefully, each simulation run will exercise a different trajectory in the state space. The next side note is very important in this regard.

Ergodic Systems

If a dynamic system is run for a long period of time, then each possible system state would be visited. Then, the mean over the state space (i.e., ensemble mean) can be approximated by the mean of a sample path through the state space (i.e., temporal mean.) Such dynamic systems are referred to as *ergodic* systems wherein the temporal mean converges to the ensemble mean.

4.5 SIMULATING QUEUEING PROCESSES

Queueing is a natural phenomenon that arises whenever there is a competition for a single resource. For example, when a patient arrives and the doctor is busy, he has to wait. This situation can be modeled using a server and a buffer. The server represents the doctor and the buffer represents the waiting area where patients sit. In this example, patients compete for a single doctor.

The total number of patients in the clinic (i.e., the ones waiting and the one being checked by the doctor) is a random variable. Since this random variable varies with time, a stochastic process emerges. This process is referred to as a queueing process. The state of this process is the total number of patients in the clinic. The queueing process changes its state whenever a new patient arrives or an existing patient leaves the clinic. The first situation is referred to as the *arrival event*. The second situation, however, is called the *departure event*.

If we let $N(t)$ be the total number of patients in the clinic at time t, then $N(t)$ is a discrete-state stochastic process. Besides, if $N(t)$ is observed at integer times (i.e., $t \in \{0, 1, 2, 3, ...\}$), then the resulting queueing process is a discrete-time process. On the other hand, if $N(t)$ is observed at random times (i.e., $t \in [0, \infty)$), then the resulting queueing process is a continuous-time process.

If $N(t)$ is a discrete-time, discrete-state process, it is referred to as a discrete-time *Markov chain*. A discrete-time Markov chain stays in any state for an amount of time which is *geometrically* distributed. In some models, the assumption is relaxed and the amount of time spent in every state is assumed to be fixed. Thus, the x-axis (i.e., time) is divided into intervals of equal lengths (also called slots). Figure 4.17 shows the behavior of a discrete-time Markov chain over nine time slots. It is assumed that events occur only at the beginning of each time slot.

On the contrary, if $N(t)$ is a continuous-time, discrete-state process, it will be referred to as a continuous-time *Markov chain*. When a continuous-time Markov chain enters a state, it remains in the state for an amount of time which is *exponentially* distributed. Figure 4.18 shows the behavior of a continuous-time Markov chain over a continuous interval of time. Events can occur at random instants of time.

Finally, it should be emphasized that whether the time is continuous or discrete, the state of a Markov chain is always discrete. Also, it should be understood that a Markov chain representing a queueing process is driven by two events: arrival and departure. When an arrival occurs, $N(t)$ is incremented by one. On the other hand, when a departure occurs, $N(t)$ is decremented by one. For the rest of this section, we are going to learn how to simulate these two stochastic processes.

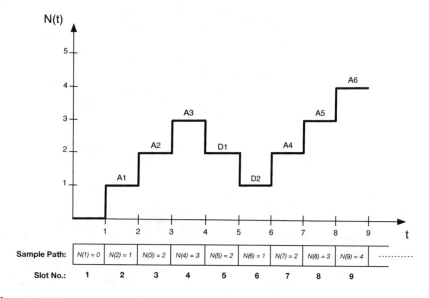

Figure 4.17

A sample path of a discrete-time Markov chain over nine time units. Events occur at integer times only. $N(1)$ is the number of entities in the system during the first time slot.

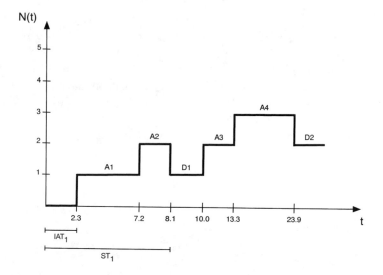

Figure 4.18

A sample path of a continuous-time Markov chain. Events occur at random times. The time spent in a state has an exponential distribution.

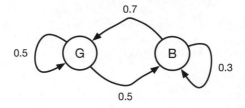

Figure 4.19
A graphical representation of a two-state, discrete-time Markov chain.

4.5.1 Discrete-Time Markov Chains

A Markov chain is said to be fully characterized if its probability transition matrix is known. An entry P_{ij} in this matrix represents the probability that the process will make a transition from state i to state j, where i is the present state and j is the next state.

Consider a Discrete-Time Markov chain (DTMC) with two states: Good (G) and Bad (B). Let the probability transition matrix of this DTMC be the following:

$$P(i,j) = \begin{matrix} & \begin{matrix} G & B \end{matrix} \\ \begin{matrix} G \\ B \end{matrix} & \begin{pmatrix} 0.5 & 0.5 \\ 0.7 & 0.3 \end{pmatrix} \end{matrix}.$$

Notice that every row and column is labeled. This matrix is an example of the classical Gilbert-Elliot two-state wireless channel model. Figure 4.19 is a graphical representation of the Markov chain, which is more understandable.

In simulating this DTMC, our objective is to generate a sequence $\{X_n, n = 1, 2, 3, ...\}$ which follows the above probability transition matrix. The subscript n is used instead of t because the time is discrete. This is just a convention. The initial state X_0 must be given before the start of the simulation.

Given that the present state is $X_n = G$, the next state X_{n+1} has the following PMF.

$$P(X_{n+1} = G) = 0.5, \qquad P(X_{n+1} = B) = 0.5.$$

Similarly, if the present state is $X_n = B$, then the next state X_{n+1} has the following PMF.

$$P(X_{n+1} = G) = 0.7, \qquad P(X_{n+1} = B) = 0.3.$$

Since we know the PMF for the next state given any present state, we can now simulate the DTMC. In fact, the task of simulating the DTMC boils down to simulating the random variable X_{n+1} as follows.

$$\text{If} \quad X_n = G, \quad X_{n+1} = \begin{cases} G, & \text{if } u \in (0, 0.5) \\ B, & \text{if } u \in [0.5, 1.0). \end{cases}$$

$$\text{If} \quad X_n = B, \quad X_{n+1} = \begin{cases} G, & \text{if } u \in (0, 0.7) \\ B, & \text{if } u \in [0.7, 1.0). \end{cases}$$

where u is a uniform random number between 0 and 1.

Listing 4.3 shows a program that generates a possible trajectory of the above DTMC given that the initial state is $X_0 = G$. The output of the program could be the following.

$$X_0 = G, \ X_1 = G, \ X_2 = B, \ X_3 = G, \dots.$$

Listing 4.3
Simulating a two-state discrete-time Markov chain given its probability transition matrix and an initial state.

```python
from random import random

n = 10
S = []

S.append('G')    # Initial state

for i in range(n):
    u = random()
    if S[i] == 'G':
        if u < 0.5:
            S.append('G')
        else:
            S.append('B')
    elif S[i] == 'B':
        if u < 0.7:
            S.append('G')
        else:
            S.append('B')

print('Sample Path: ', S)
```

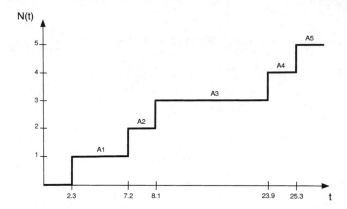

Figure 4.20
Sample path of a Poisson process. Only arrival events occur inside a Poisson process.

4.5.2 Continuous-Time Markov Chains

A Poisson process is an example of a Continuous-Time Markov Chain (CTMC). You can think of it as a counter which counts the events which have occurred so far. The time between two consecutive events is called the *Inter-Arrival Time* (IAT). The random variable IAT has an exponential distribution. Only one kind of event triggers a transition inside a Poisson process. This event is the arrival event. Figure 4.20 shows one possible sample path of a Poisson process. Listing 4.4 shows how a Poisson process can be simulated in Python. Notice the variable defined on line 6. It is used to keep track of the simulation time. Also, notice how the simulation time is advanced on line 11. Basically, the time of the next arrival event is equal to the current simulation time plus the IAT. The simulation loop (lines 8-11) is terminated once the current simulation time exceeds preset total simulation time.

```
Listing 4.4
Simulating a Poisson process.

1   from random import expovariate

2

3   Avg_IAT = 2.0      # Average IAT

4   Sim_Time = 100     # Total simulation time

5   N = 0              # Count number of arrivals

6   clock = 0          # Simulation time

7
```

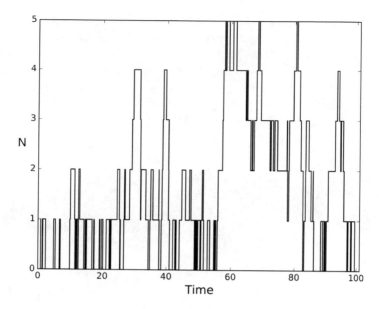

Figure 4.21
Sample path of a birth-death process.

```python
8   while clock <= Sim_Time:

9       N = N + 1

10      # Advance simulation clock

11      clock = clock + expovariate(1/Avg_IAT)

12

13  print('Total Number of Arrivals = ', N)
```

The Poisson process is a special case of another type of random processes called Birth-Death (BD) processes. In BD process, two events occur: birth and death. The Poisson process is a pure birth process since the birth (i.e., arrival) event occurs only. The state of a BD process changes at random points of time. The state variable is incremented by one when a birth event occurs. It is decremented by one, on the other hand, when a death occurs. The time until the next birth is exponentially distributed with rate λ. Similarly, the time until the next death is exponentially distributed with rate μ.

A BD process is used to model the number of customers in the single-server queueing system.[5] It can be simulated as shown in Listing 4.5. In each

---

[5] In the terminology of queueing theory, this system is referred to as an M/M/1 queueing system.

iteration of the simulation loop (lines 12-25), two exponentially distributed random *variates* (i.e., numbers) are generated. These two numbers represent the inter-arrival times for the next birth and death events. The smaller of the two numbers is the time at which the next event occurs. Figure 4.21 shows a possible sample path of a birth-death process. This figure is generated using lines 27-32 in Listing 4.5. Pay attention to how data is collected on lines 18-19 and 24-25. The lists used for data collection are defined as empty lists on lines 9-10.

Listing 4.5
Simulating a birth-death process and plotting its sample path (see Figure 4.21).

```python
1   from random import expovariate
2   from matplotlib.pyplot import *
3
4   Avg_IAT = 2.0
5   Avg_ST = 1.0     # Avg service time
6   Sim_Time = 100   # Total simulation time
7   N = 0
8   clock = 0    # Simulation time
9   X = []   # Times of events
10  Y = []   # Values of N
11
12  while clock <= Sim_Time:
13      IAT = expovariate(1 / Avg_IAT)
14      ST = expovariate(1 / Avg_ST)
15      if IAT <= ST:
16          N += 1
17          clock = clock + IAT
18          X.append(clock)
19          Y.append(N)
20      else:
21          if N > 0:
22              N -= 1
23              clock = clock + ST
24              X.append(clock)
```

```
25              Y.append(N)

26

27   step(X, Y, Linewidth=1.2, color='black')

28   xlabel('Time', size=16)

29   ylabel('N', size=16)

30   xlim(0, 101)

31   #show()

32   savefig('sim_birth_death_process.pdf', format='pdf',
          bbox_inches='tight')
```

4.6 SUMMARY

In this chapter, you have learned about several important random variables and their probability distribution functions. You have also learned about stochastic processes and their fundamental role in system modeling. In addition, you have learned new conventions when writing simulation programs in Python. For example, you should be comfortable now with the following programming concepts:

1. Simulation loop,

2. Keeping track of simulation time using the *clock* variable,

3. Advancing the simulation time using randomly generated numbers, and

4. Using lists for collecting simulated data.

You will need all these concepts and techniques in the next chapters.

4.7 EXERCISES

4.1 Write a Python program to plot the PDF and CDF of the Erlang random variable.

4.2 A Bernoulli random process $X(n)$ counts the number of successes at the end of the nth time slot. Let the initial state be $X(0) = 0$. Write a Python program which simulates this process over 15 time slots. Plot one sample path.

4.3 A manufacturer distributes a coupon in every box he makes. The coupon put in each box is chosen randomly from a set of N distinct coupons. Your goal is to collect all the N distinct coupons. Write a Python program to estimate the expected number of boxes that you must buy.

4.4 Random walk is a discrete-time, discrete-state stochastic process. Let the state of the process represent the direction of movement: North, South, East, and West. Write a Python program which simulates this process. Assume that the probabilities of moving North, South, East, and West are 0.1, 0.5, 0.3, and 0.1, respectively. Assume a simulation time of 15 time units. Plot one sample path of the process.

## CHAPTER 5

Simulating the Single-Server Queueing System

"Learning by doing and computer simulation are all part of the same equation."
—Nicholas Negroponte

The single-server queueing system is very fundamental. A complex system such as the Internet can be decomposed into a set of interacting single-server queueing systems. In this chapter, we are going to study in detail how to simulate the single-server queueing system. In addition, we are going to learn about several performance laws which can be used to assess the performance of computer and network systems. Further, the process of collecting simulated data to be used in computing performance measures is described. Performance measures are computed as averages of multiple simulation runs. This is why the method of independent replications is also discussed in this chapter. After that, two methods are described for determining the length of the transient phase in a simulation run. Finally, manual simulation is discussed to reinforce the above topics.

5.1 SIMULATION MODEL

Before a simulation model of a system can be constructed, we need to understand the physical structure of the system. Figure 5.1 shows the physical structure of the single-server queueing system. This system has four components: source, buffer, server, and sink. The figure also shows how these components are connected. Basically, the source generates packets which go

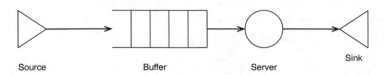

Source   Buffer   Server

Figure 5.1
Physical structure of the single-server queueing system.

into a buffer. The server fetches the packets from the buffer and then delivers them to the sink after they are processed.

Packets are transferred to the server in the same order in which they enter the buffer. This buffering mechanism is referred to as the First-In First-Out (FIFO) mechanism. This observation is very helpful when collecting simulated data. That is, since the order of packets is maintained by a FIFO policy, there is no need to assign indexes (or identifiers) to packets. The first packet which enters the system is going to be the first packet which leaves the system. The same observation applies to all the subsequent packets.

Since the individual inter-arrival times and service times are unpredictable, they are modeled as random variables. Thus, we need to specify the probability distributions of these two random variables. The choice of a specific probability distribution has to be supported by an evidence that it is appropriate. The exponential probability distribution is a reasonable model of the inter-arrival and service times.

Listing 5.1 shows a Python implementation of the simulation model of the single-server queueing system. It is based on the C-language implementation provided in [8]. In this simulation model, there are two fundamental events: arrival and departure. The state variable N represents the number of packets inside the system. It is the state of the random process we are going to observe. Remember that this process is a BD process. The birth and death events are the arrival and departure events of a packet, respectively.

The arrival process is a Poisson process with an average inter-arrival time of 2.0 (line 4). The departure process is also a Poisson process with an average service time of 1.0 (line 5). The system will be simulated for 100.0 time units (line 6). The simulation clock is initialized to zero and it is used to keep track of the simulation time (line 7).

For every event, a variable is needed to keep track of its time of occurrence. For the arrival event, the variable Arr_Time is used. After an arrival occurs, this variable is updated with the time of the next arrival. Similarly, the variable Dep_Time keeps track of the time of next departure (i.e., service completion). The variable clock represents the current simulation time. It acts like an internal clock for the simulation model.

If the system becomes empty (i.e., $N = 0$) due to a departure event, the variable Dep_Time is set to ∞ to ensure that the next event will be an arrival. This is also done in the initialization phase to ensure that the first event will

Listing 5.1
Simulation program of the single-server queueing system.

```python
from random import expovariate
from math import inf as Infinity

Avg_IAT = 2.0          # Average Inter-Arrival Time
Avg_ST = 1.0           # Average Service Time
Tot_Sim_Time = 100.0   # Total Simulation Time
clock = 0.0            # Current Simulation Time

N = 0 # State variable; number of customers in the system

# Time of the next arrival event
Arr_Time = expovariate(1.0/Avg_IAT)
# Time of the next departure event
Dep_Time = Infinity

while clock <= Tot_Sim_Time:
  if Arr_Time < Dep_Time:   # Arrival Event
    clock = Arr_Time
    N = N + 1.0
    Arr_Time = clock + expovariate(1.0/Avg_IAT)
    if N == 1:
      Dep_Time = clock + expovariate(1.0/Avg_ST)
  else:                     # Departure Event
    clock = Dep_Time
    N = N - 1.0
    if N > 0:
      Dep_Time = clock + expovariate(1.0/Avg_ST)
    else:
      Dep_Time = Infinity
```

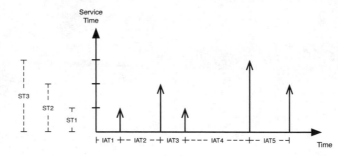

(a) Inter-arrival times and service times for five packets.

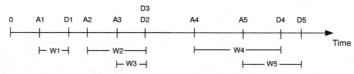

(b) Arrival and departure events mapped onto the time line. For each packet i, W_i represents the total time spent in the system.

Figure 5.2
Graphical representation of the relationship between random variables and simulation events.

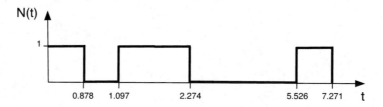

Figure 5.3
A sample path of the random process $N(t)$.

be an arrival. In short, you cannot have a departure while $N = 0$. Note that multiple arrival events may occur before a departure event.

Figure 5.3 illustrates the behavior of the simulated system. The first departure occurs at time 0.878. This is why there is a line segment extending from time 0.0 to time 0.878. The graph of $N(t)$ is a series of line segments. The length of every line segment represents the time until the next event.

Figure 5.4(a) shows that three random processes can be defined in the single-server queueing system: arrival, departure, and queueing. Figures 5.4(b), (c), and (d) show sample paths of these three random processes. Notice that the departure process is a shifted version of the arrival process. Also, notice how the behavior of the queueing process is defined as the result of the interaction between the arrival and departure processes.

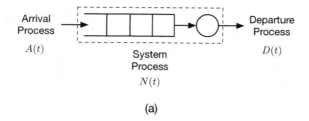

(a)

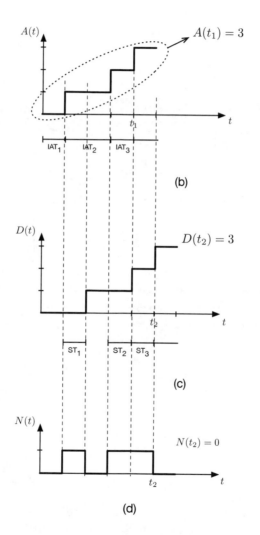

Figure 5.4
Random processes present in the single-server queueing system. Both the arrival and departure processes are Poisson processes. (a) Places where the random processes are defined. (b) Total number of arrivals which have occurred up to time t_1 is three. (c) The sample path of the departure process is a shifted version of the sample path of the arrival process. (d) Sample path of the queueing (birth-death) process which tracks the number of packets in the system.

What Causes Delay?

When multiple packets contend for one server, some packets will be queued and system performance suffers. If the service time is always less than or equal to the inter-arrival time, no packet is queued. In reality, however, the service times and inter-arrival times are not constant. Also, packets may require different service times. This variability in service times and inter-arrival times causes the delay through the single-server queueing system.

Table 5.1
Manual simulation of the single-server queueing system using a simulation table.

Pkt No.	IAT	ST	Arrival Time	Service Starts At	Departure Time	Time in Queue	Time in System
1	2	12	2	2	14	0	12
2	5	10	7	14	24	7	17
3	1	16					
4	4	9					
5	1	10					
6	3	13					
7	3	17					
8	2	10					
9	4	8					
10	5	12					

Manual Simulation

Table 5.1 shows how a manual simulation of the single-server queueing system shown in Figure 5.1 can be performed. The simulation table has eight columns, which are divided into two groups. The first three columns represent the information needed before starting the simulation. The fourth column is used to record the absolute arrival time which is $clock + IAT$. The fifth column is the time at which the service of a packet starts. The service of packet number i starts at the time of departure of packet number $i-1$. Of course, the service of the first packet starts immediately. The departure time is recorded in the sixth column. The waiting time in the queue is the difference between the departure time and arrival time. It is captured in the seventh column. The last column is used for recording the total time a packet spends in the system (i.e., system response time).

5.2   COLLECTING SIMULATED DATA

The program in Listing 5.1 is not of much use until we add *output variables* to it. Output variables are lists and dictionaries used for collecting simulated data which is then used in computing performance measures. In this section, the notion of output variables is motivated. Several examples will be given in the next section to show how output variables are used in simulation programs.

As shown in Figure 5.5(a), a simulation experiment is composed of a set of input variables, parameters, output variables, and simulation model. The simulation model is not shown explicitly. The model transforms the values of the input variables into new values constituting the output variables. Values of the input variables can be generated while the simulation program is running. They can also be specified at the beginning of a simulation run. A simulation run may mean fixing the parameters and changing the values of the input variables. The opposite is also possible.

As an example, consider a simulation experiment for estimating the average response time of the single-server queueing system. The response time is the delay a packet experiences while traveling through the sytem. Figure 5.5(b) shows the setup necessary to perform this experiment. In this setup, there are two input variables: IAT and ST. Each input variable represents a sequence of random numbers. These random numbers are generated using appropriate probability distributions, which are parametrized by λ and μ. There is only one output variable, which is Delay. The third parameter Num_Pkts represents the length of each simulation run (i.e., number of packets to be simulated).

Each entry in the output variable Delay corresponds to the delay of one simulated packet. It is the result of subtracting the arrival time from the departure time for each packet upon leaving the system. At the end of the simulation run, Delay will contain n observations (or samples).

$$Delay = [d_1, d_2, d_3, ..., d_n],$$

where d_i is the delay experienced by the ith simulated packet. It should be pointed out that each sample d_i is a random variable. This is because if the same simulation experiment is repeated again, the value of d_i will be different. As a result, Delay is also a random variable with an unknown probability distribution.

Why Output Variables?

A simulation run results in an output variable whose expected value is an estimate of the performance measure of interest. For example, for the output variable Delay, statistics.mean(Delay) gives the estimate of the performance measure. The values in the list referred to by Delay represent a sample path of the single-server queueing system.

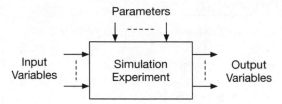

(a) Elements of a simulation experiment. Simulation model is hidden.

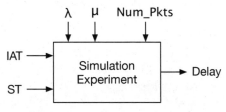

(b) Single-server queueing system.

Figure 5.5
A simulation experiment represents an execution of a simulation model with a specific set of parameters, inputs, and outputs.

5.3   PERFORMANCE LAWS

Let the variable `Tot_Sim_Time` in Listing 5.1 be represented by T. So, T becomes the length of the period of time over which the system is simulated. Also, let D denote the number of departures that occur during a simulation run of length T. The following fundamental laws can be used to measure the performance of the single-server queueing system.

5.3.1   Throughput

Throughput measures how many packets the system can process in one time unit. It is defined as the ratio of the number of departures divided by the total simulation time. Mathematically, this law can be written as follows.

$$\tau = \frac{D}{T}. \tag{5.1}$$

The unit of throughput is packets per a time unit (pkt/time unit).

5.3.2   Utilization

Server utilization is the proportion of simulation time during which the server is busy. It is the product of its throughput and the average service time per customer. This can mathematically be expressed as follows.

$$U = \tau \cdot T_s \tag{5.2}$$

where T_s is the average service time per customer and it is defined as follows.

$$T_s = \frac{B}{D} \tag{5.3}$$

where B is the total server busy time which can be computed as follows.

$$B = \sum_{i=1}^{D} T_i \tag{5.4}$$

where T_i is the service time for customer i.

5.3.3 Response Time

The response time is the total time a customer spends in the system. That includes waiting time (or queueing time) and service time. Another name for the response time is *delay*.

Define W_i as the time spent in the system by the ith simulated packet. Then, the average response time of the system can be computed as follows.

$$W = \frac{\sum_{i=1}^{D} W_i}{D}. \tag{5.5}$$

As a consequence, the average number of packets in the system can be computed as follows.

$$L = \tau \cdot W. \tag{5.6}$$

Listing 5.2
Estimating the average response time of the system.

```
1  from random import expovariate
2  from statistics import mean
3  from math import inf as Infinity
4
5  # Parameters
6  lamda = 1.3          # Arrival rate (Lambda)
7  mu = 2.0             # Departure rate (Mu)
8  Num_Pkts = 100000    # Number of Packets to be simulated
9  count = 0      # Count number of simulated packets
```

```python
10  clock = 0
11  N = 0           # State Variable; number of packets in system
12
13  Arr_Time = expovariate(lamda)
14  Dep_Time = Infinity
15
16  # Output Variables
17  Arr_Time_Data = []    # Collect arrival times
18  Dep_Time_Data = []    # Collect departure times
19  Delay_Data = []    # Collect delays of individual packets
20
21  while count < Num_Pkts:
22    if Arr_Time < Dep_Time:    # Arrival Event
23      clock = Arr_Time
24      Arr_Time_Data.append(clock)
25      N = N + 1.0
26      Arr_Time = clock + expovariate(lamda)
27      if N == 1:
28        Dep_Time = clock + expovariate(mu)
29    else:                # Departure Event
30      clock = Dep_Time
31      Dep_Time_Data.append(clock)
32      N = N - 1.0
33      count = count + 1 # Packet Simulated
34      if N > 0:
35        Dep_Time = clock + expovariate(mu)
36      else:
37        Dep_Time = Infinity
38
39  for i in range(Num_Pkts):
40    d = Dep_Time_Data[i] - Arr_Time_Data[i]
41    Delay_Data.append(d)
42
43  print( "Average Delay = ", round( mean(Delay_Data), 4) )
```

> **Little's Law**
>
> $$L = \lambda \cdot W$$
>
> This law asserts that the time average number of packets in the system is the product of the arrival rate and the response time. This law is due to Little who proved it in 1961 [7]. Remember that λ is a parameter of the arrival Poisson process. In simulation, it is the argument passed to the function `random.expovariate`.

Listing 5.2 shows the Python code necessary to perform the experiment in Figure 5.5(b). In this program, the values of the input variables are generated whenever they are needed. On the other hand, for the output variable `Delay`, a list is explicitly defined to hold its values. These values are the results of subtracting the values in two intermediate output variables (i.e., `Arr_Time_Data` and `Dep_Time_Data`). At the end of the program, the mean function in the `statistics` module is applied on the `Delay` output variable to get the average of the individual packet delays.

5.3.4 $\mathbb{E}[N(t)]$

The state variable $N(t)$ represents the number of packets in the system at time t. In the previous section, the Little's law is used to compute the average number of customers in the system; i.e., $\mathbb{E}[N(t)]$. This quantity can be directly computed by using one sample path of $N(t)$ as follows:

$$\mathbb{E}[N(t)] = \frac{1}{T} \cdot \int_0^T N(t), \qquad (5.7)$$

where T is the total simulation time.

The integral in Eqn. (5.7) is the sum of the areas of the individual rectangles under the curve of $N(t)$. For example, in Figure 5.6, there are eight rectangles. The length of each rectangle is equal to the number of packets in the system while the width is the time interval between the events causing the change in N. Hence, the areas of the eight rectangles are 0, 2, 2, 3, 6, 1, 0, and 1. Since the total simulation time is 12, then the average number of packets in the system can be computed as follows.

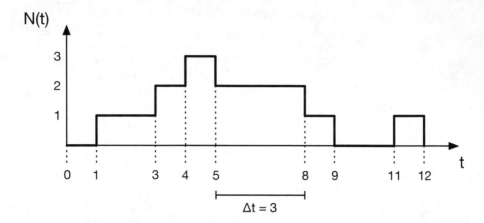

Figure 5.6
A sample path of the number of packets in the single-server queueing system. There are eight rectangles under the curve of the sample path.

$$\mathbb{E}[N(t)] = \frac{0+2+2+3+6+1+0+1}{12}$$
$$= \frac{15}{12}$$
$$= 1.25.$$

Listing 5.3 shows how the above technique can be implemented in Python. The new code is on lines 16, 17, 22-24, 31-33, and 41. A new variable is defined on line 16 and it is used to record the time of occurrence of the last simulated event. Inside the simulation loop, after updating the simulation clock, the area of the current rectangle delimited by the current and previous events is calculated on lines 23 and 32. After this operation, the value of the variable `Prev_Event_Time` is changed to the current simulation time. At the end of the simulation run, the total area under the curve is computed as shown on line 41. The variable `clock` stores the total simulation time.

> **Listing 5.3**
> Estimating the average number of customers in the sytem ($\mathbb{E}[N(t)]$).

```
1   from random import expovariate
2   from statistics import mean
3   from math import inf as Infinity
4
5   # Parameters
```

```
 6  lamda = 1.3
 7  mu = 2.0
 8  Num_Pkts = 1000000
 9  count = 0
10  clock = 0
11  N = 0
12
13  Arr_Time = expovariate(lamda)
14  Dep_Time = Infinity
15
16  Prev_Event_Time = 0.0 # Time of last event
17  Area = [] # Output variable
18
19  while count < Num_Pkts:
20    if Arr_Time < Dep_Time:
21      clock = Arr_Time
22      # Area of rectangle
23      Area.append((clock - Prev_Event_Time) * N)
24      Prev_Event_Time = clock
25      N = N + 1.0
26      Arr_Time = clock + expovariate(lamda)
27      if N == 1:
28        Dep_Time = clock + expovariate(mu)
29    else:
30      clock = Dep_Time
31      # Area of rectangle
32      Area.append((clock - Prev_Event_Time) * N)
33      Prev_Event_Time = clock
34      N = N - 1.0
35      count = count + 1
36      if N > 0:
37        Dep_Time = clock + expovariate(mu)
38      else:
39        Dep_Time = Infinity
```

```
40
41  print( "E[ N(t) ] = ", round(sum(Area) / clock, 4) )
```

5.3.5 $\mathbb{P}[N]$

$\mathbb{P}[N = k]$ is the probability that there are exactly k packets in the system. In order to estimate this probability, we sum up all time intervals during which there are exactly k packets in the system. Then, the sum is divided by the total simulation time. For instance, in Figure 5.6, the system contains one packet only during the following intervals: $[1, 3]$, $[8, 9]$, and $[11, 12]$. Thus, the probability that there is exactly one packet in the system can be estimated as follows.

$$\mathbb{P}[N = 1] = \frac{(3 - 1) + (9 - 8) + (12 - 11)}{12}$$
$$= \frac{2 + 1 + 1}{12}$$
$$= 0.33.$$

Listing 5.4 shows how $\mathbb{P}[N = k]$ can be estimated using simulation. The new code is on lines 15, 17, 22-26, 33-37, 45-47, and 49-50. In this program, a new data structure called *dictionary* is used. In a dictionary, keys are used for storing and fetching items. A dictionary is defined using two curly braces as shown on line 17. This defines an empty dictionary. The dictionary is populated on lines 24-25 and 35-36. Basically, if the key N is already used, the value which corresponds to this key is updated. Otherwise, a new key is inserted into the dictionary and its value is initialized. The value of the key is updated using the length of the current time interval on lines 23 and 34. As in the previous example, the time of the current event is saved to be used in the next iteration of the simulation loop. Also, the state variable N is updated after computing the time interval and updating the dictionary.

In order to verify the simulation program, two checks are performed. First, on lines 49-50, the sum of probabilities is checked to be equal to one. Second, on lines 52-55, the mean is computed and compared against the theoretical value. If the two checks evaluate to true, then the simulation program is correct.

Listing 5.4
Estimating the steady-state probability distribution ($\mathbb{P}[N = k]$).

```
1  from random import expovariate
2  from statistics import mean
```

```python
3   from math import inf as Infinity
4
5   # Parameters
6   lamda = 1.3
7   mu = 2.0
8   Num_Pkts = 1000000
9   count = 0
10  clock = 0
11  N = 0
12
13  Arr_Time = expovariate(lamda)
14  Dep_Time = Infinity
15  Prev_Event_Time = 0.0
16
17  Data = {} # Dictionary
18
19  while count < Num_Pkts:
20    if Arr_Time < Dep_Time:
21      clock = Arr_Time
22      # Length of time interval
23      delta = clock - Prev_Event_Time
24      if N in Data: Data[N] += delta
25      else:       Data[N] = delta
26      Prev_Event_Time = clock
27      N = N + 1.0
28      Arr_Time = clock + expovariate(lamda)
29      if N == 1:
30        Dep_Time = clock + expovariate(mu)
31    else:
32      clock = Dep_Time
33      # Length of time interval
34      delta = clock - Prev_Event_Time
35      if N in Data: Data[N] += delta
36      else:       Data[N] = delta
```

```
37      Prev_Event_Time = clock
38      N = N - 1.0
39      count = count + 1
40      if N > 0:
41         Dep_Time = clock + expovariate(mu)
42      else:
43         Dep_Time = Infinity
44
45   # Compute probabilities
46   for (key, value) in Data.items():
47     Data[key] = value / clock
48
49   # Check total probability is 1.0
50   print("Sum of Prob's = ", sum( Data.values() ) )
51
52   # Check expectation
53   mean = 0.0
54   for (key, value) in Data.items():
55     mean = mean + key * value
56
57   print("E[N] = ", mean)
```

5.4   INDEPENDENT SIMULATION RUNS

Figure 5.7 shows the raw data which is generated as a result of running a simulation program. Each simulation run results in an instance of the output variable. Each realization of the output variable is a vector containing n observations. Each instance of the output variable accounts for one sample of the performance measure. The first k observations in the output variable are dropped because they are part of the transient phase. In this phase, data is highly variable due to the initial conditions. In the single-server example, for instance, the initial conditions are that the system is empty and the server is idle.

In an introductory statistics course, we always assume that samples in a data set are IID. In fact, you should know that classical statistical techniques can be applied only on data sets with IID samples. Unfortunately, in a simu-

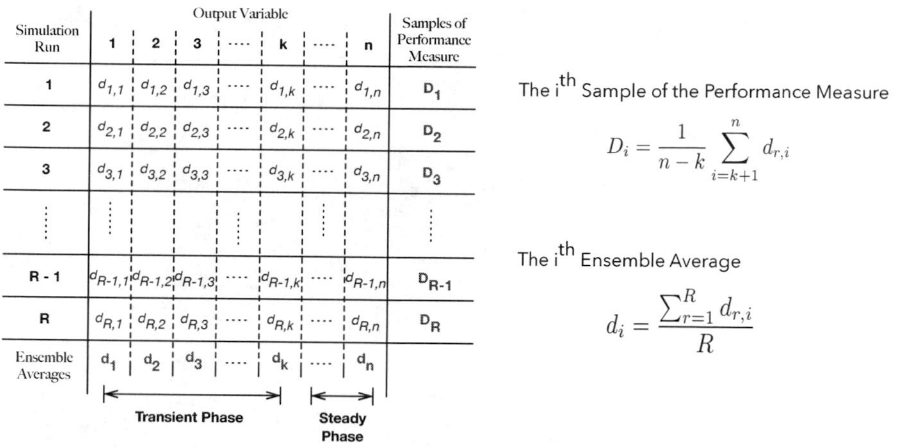

Simulation Run	Output Variable						Samples of Performance Measure
	1	2	3	…. k	….	n	
1	$d_{1,1}$	$d_{1,2}$	$d_{1,3}$	…. $d_{1,k}$	….	$d_{1,n}$	D_1
2	$d_{2,1}$	$d_{2,2}$	$d_{2,3}$	…. $d_{2,k}$	….	$d_{2,n}$	D_2
3	$d_{3,1}$	$d_{3,2}$	$d_{3,3}$	…. $d_{3,k}$	….	$d_{3,n}$	D_3
⋮	⋮	⋮	⋮	⋮	⋮	⋮	⋮
R - 1	$d_{R-1,1}$	$d_{R-1,2}$	$d_{R-1,3}$	…. $d_{R-1,k}$	….	$d_{R-1,n}$	D_{R-1}
R	$d_{R,1}$	$d_{R,2}$	$d_{R,3}$	…. $d_{R,k}$	….	$d_{R,n}$	D_R
Ensemble Averages	d_1	d_2	d_3	…. d_k	….	d_n	

Transient Phase | Steady Phase

The ith Sample of the Performance Measure

$$D_i = \frac{1}{n-k} \sum_{i=k+1}^{n} d_{r,i}$$

The ith Ensemble Average

$$d_i = \frac{\sum_{r=1}^{R} d_{r,i}}{R}$$

Figure 5.7
Raw data generated when running a simulation program.

lation run, values of an output variable are not independent. Thus, the value of a performance metric resulting from a single simulation run cannot be used as an estimate.

For example, in order to compute an estimate of the response time (W) of the single-server queueing system, we need to define an output variable (say) $Z = [W_i]$, where Z is a list of the times each simulated packet spends in the system. The response time for each packet is computed as $W_i = T_{q_i} + T_{s_i}$, where T_{q_i} is the time spent in the queue and T_{s_i} is the time spent at the server. Note that $W_i = W_{i-1} + W_{i-2} + ... + W_1$. Hence, we can conclude that W_i's are not independent. This serial dependence exists in both the transient and steady phase. So, what should we do?

The simplest remedy to the above problem is to construct your sample set by making multiple independent simulation runs. In this case, each simulation run will generate one sample in your sample set. In this way, you will have a sample set with IID samples and thus can apply the classical statistical techniques. Listing 5.5 shows how you generate multiple independent samples of the delay performance measure using the simulation model of the single-server queueing system defined in the external library simLib.

The number of independent simulation runs to be performed is stored in the variable Num_Repl. In each simulation run, n packets are simulated. The former is necessary to ensure IID samples. Also, the latter is necessary to ensure that the transient phase is eliminated.

Finally, to make sure that every simulation run is independent from all the other simulation runs, you have to *reseed* the random number generator (see line 17). That is, for every simulation run, you must assign a unique *seed* to the function random(). This way the sequence of generated random numbers will be different. This approach should result in independent samples of your performance measure. This requirement will become clear in Chapter 9 when we discuss random number generators.

Listing 5.5
Performing multiple independent simulation runs of the simulation model of the single-server queueing system.

```python
# simLib is your simulation library, which you will reuse
# in your homework and projects.
# It is available in the github repository

from simLib import mm1
from random import seed
from statistics import mean

lamda = 1.3
mu = 2
n = 100000   # Number of packets to be simulated

Num_Repl = 50    # Number of replications (repetitions)
Delay = []        # Data set

for i in range(Num_Repl):
    seed()   # Reseed RNG
    d = mm1(lamda, mu, n)
    Delay.append(d)

# Estimate of performance measure
print("Average Delay = " , round( mean(Delay), 4) )
```

5.5   TRANSIENT AND STEADY PHASES

Figures 5.8(a) and 5.8(b) show that a simulation run goes through two phases: *transient* and *steady*. In the transient phase , the values of the output variable

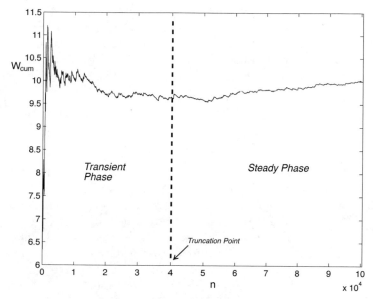

(a) As the number of simulated packets increases, the cumulative average approaches the theoretical value. The truncation point is $n = 40000$.

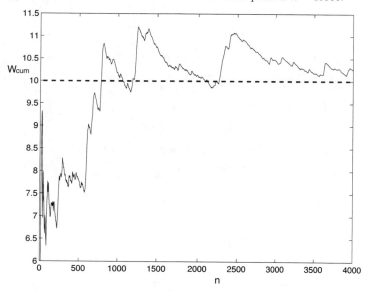

(b) This initial interval is considered to be part of the transient phase of the simulation.

Figure 5.8
Cumulative average versus number of simulated packets. The theoretical value is $W_{avg} = 10$. After the transient phase is over, the cumulative average starts approaching the theoretical value.

W_{cum} vary dramatically. They are significantly different from the theoretical value ($W_{avg} = 10$) computed using standard queueing theory formulas. W_{cum} can be computed using Eqn. (5.8), where n is the number of simulated packets. Finally, in this specific simulation run, the transient phase extends from one to approximately $n = 40000$ simulated packets. That means the first 40000 samples are dropped from the output variable. At the end of the simulation run, the output variable will contain only 60000 samples.

$$W_{cum} = \frac{\sum_{i=1}^{n} W_i}{n}. \tag{5.8}$$

In the transient phase, output variables fluctuate due to the effect of the initial state of the simulation model. Thus, no simulation data should be collected during this phase. Instead, the simulation program should be allowed to run until it exits this phase. Interestingly, this phase is also referred to as the warm-up phase. Figure 5.8(b) shows a detailed view of the transient phase.

Several techniques exist for estimating the length of the transient phase . In this book, we are going to use a simple but effective technique based on the Welch's method introduced in [11]. This technique uses the running average of the output variable. Several realizations of the output variable are generated. Then, they are combined into one sequence in which each entry represents the average of the corresponding entries in the generated realizations. This final sequence is then visually inspected to identify an appropriate truncation point. Figure 5.9 shows the final sequence (Z) resulting from five realizations. The entries before the truncation point will be discarded. This is because they will introduce a bias in the point estimate of the performance measure.

The first two steps in the Welch's method are shown in Figure 5.10. The following is a description of these two steps.

1. For each output variable Y, run the simulation at least five times. Each simulation run i generates a realization $Y[i]$ of size m.

2. Calculate the mean across all the generated realizations, i.e.,

$$Z[i] = \frac{\sum_{i=1}^{R} Y[i]}{R}, \tag{5.9}$$

where R is the number of simulation runs performed in step 1.

3. Plot the sequence Z.

4. The warm-up period ends at a point k when the curve of Z becomes flat. Choose this point as your truncation point.

Listing 5.6 gives a Python implementation of the above two-step technique for determining truncation points. It also gives the code used for generating Figure 5.9. As you can tell from this figure, using the average of multiple realizations is more effective than using a single realization of the output variable to determine the length of the transient phase.

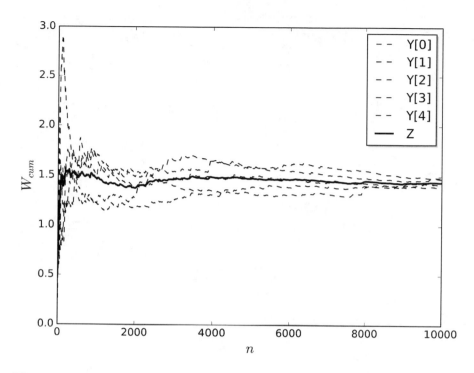

Figure 5.9
Z is the average of the five output sequences Y[0]-Y[4]. A truncation point
can visually be determined by using the curve of Z. In this example, a good
truncation point is $n = 3000$.

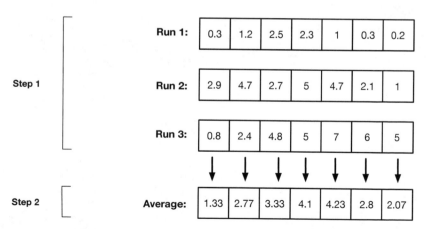

Figure 5.10
The first two steps in the Welch's method. In step 1, multiple realizations of
the output variable are generated. These realizations are combined into one
sequence in step 2.

Listing 5.6
Determining a good trunction point using the average of several realizations of an output variable.

```python
from simLib import out_var_cum_mm1
from random import seed
from matplotlib.pyplot import *
import numpy as np


lamda = 1.3
mu = 2

n = 10000 # Number of packets to be simulated
R = 5     # Number of replications (repetitions)

Y =  np.zeros( shape = (R, n) ) # Output variable Delay

# 1. Generate sample paths
for i in range(R):
    seed()
    Y[i] = out_var_cum_mm1(lamda, mu, n)

# 2. Compute the mean
Z = []
for i in range(n):
    Z.append( sum(Y[:, i]) / R )

# Plot Y and Z
plot(Y[0], "k--", label="Y[0]")
plot(Y[1], "k--", label="Y[1]")
plot(Y[2], "k--", label="Y[2]")
plot(Y[3], "k--", label="Y[3]")
plot(Y[4], "k--", label="Y[4]")
plot(Z, "k", linewidth=2, label="Z")

```

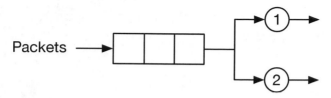

Figure 5.11
A two-server queueing system with a finite buffer of size three.

Table 5.2
IATs and STs for Exercise 5.1.

Pkt	1	2	3	4	5	6	7	8	9	10	11	12	13	14	15
IAT	2	5	1	3	1	3	3	2	4	5	3	1	1	1	2
ST	12	10	16	9	10	13	17	10	8	12	6	5	4	3	3

```
32  xlabel("$n$", size=16)
33  ylabel("$W_{cum}$", size=16)
34  legend(loc='upper right', shadow=True)
35  show()
```

5.6 SUMMARY

The selection of the next event by direct comparison of event occurrence times becomes cumbersome as the number of servers increases. In Chapter 7, you will learn about the *event list*, which is the preferred way for next event selection. This data structure is natively supported by Python and leads to concise and more manageable simulation programs. You will just need to learn how to include it in your simulation program and use it.

5.7 EXERCISES

5.1 Consider the two-server queuing system shown in Figure 5.11. The two servers are indexed from 1 to 2 and the buffer has a finite buffer of size three. That is, at most three packets can be stored inside the system at any instant of time. The Inter-Arrival Times (IATs) and Service Times (STs) for 15 packets are given in Table 5.2. A packet goes to the server with the lowest index. If all the two servers are occupied, the packet waits in the queue. Perform a manual simulation of the system and then answer the following questions:

    a. How many customers have to wait in the queue?

    b. What is the average waiting time for a packet that has to join the queue?

    c. What is the average time a packet spends in the system?

    d. What is the total utilization (i.e., busy) time for each server?

    e. What is the probability that an arriving packet is lost?

5.2 Extend the simulation program in 5.1 to simulate the system in Figure 5.11. Verify the simulation program by applying the workload (i.e., 15 packets) in Exercise 5.1 and then comparing the results with those obtained manually.

## CHAPTER 6

Statistical Analysis of Simulated Data

"The population mean is an unknown constant and no probability statement concerning its value may be made."
−Jerzy Neyman, inventor of confidence intervals.

Simulation is performed in order to generate sample paths of the system under study. Then, these sample paths are used for computing time averages of several performance measures. For each performance measure, two estimates are generally of interest: the point estimate (i.e., mean) and the interval estimate (i.e., confidence interval). It is important that a confidence interval for each mean is reported since simulation generates a finite sample path, which is also one of many possible sample paths. In this chapter, you will learn how to construct and interpret confidence intervals. You will also learn about one method for comparing two system designs using simulation.

6.1 POPULATIONS AND SAMPLES

In this section, you will be introduced to the notion of a *population* and the idea of *sampling* from that population. These two concepts are very important because statistics is the science of making inferences about the population using facts derived from random samples. To illustrate this point, consider the problem of estimating the average delay through the single-server queueing system by using simulation. In this case, we can consider a population of packets that will travel through the system. However, it would be more interesting if we consider the set of all possible delays as our population (see Figure 6.1). This will lead to a population of numbers rather than a population of individual packets.

It should be pointed out that in some applications, populations of scalars

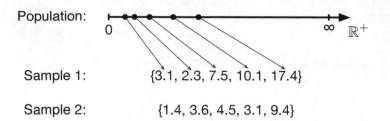

Figure 6.1
Population and samples for the simulation experiment of estimating the delay through the single-server queueing system by simulating five packets. The population is $(0, \infty)$.

(i.e., single numbers) are not sufficient. We need to consider populations of vectors. For instance, if you want to estimate the delay and throughput of the single-server queueing system, you will end up with a population of ordered-pairs (W, τ), where $W \in (0, \infty)$ and $\tau \in (0, \infty)$.

Random Samples

Each observation (or sample) of a performance metric is a random variable. Hence, a random sample of size n consists of n random variables such that the random variables are independent and have the same probability distribution. For example, in Figure 6.1, each random sample contains five observations. The first observation is different in both sample sets. This is because the first observation is a random variable and we cannot predict its value in each sample set.

Statistics, such as the sample mean and variance, are computed as functions of the elements of a random sample. Statistics are functions of random variables. The following are some of the most commonly used statistics.

1. Sample mean

$$\overline{X} = \frac{1}{n} \sum_{i=1}^{n} X_i,$$

2. Sample variance

$$S^2 = \frac{1}{n-1} \sum_{i=1}^{n} (X_i - \overline{X})^2,$$

3. Sample standard deviation

$$S = \sqrt{S^2}.$$

Table 6.1
Notation for the sample and population statistics.

	Mean	Variance	Standard Deviation	
Sample	$\overline{X}$	S^2	S	- Random variables - Obtained from simulation
Population	μ	σ^2	σ	- Constants - Obtained from queueing theory

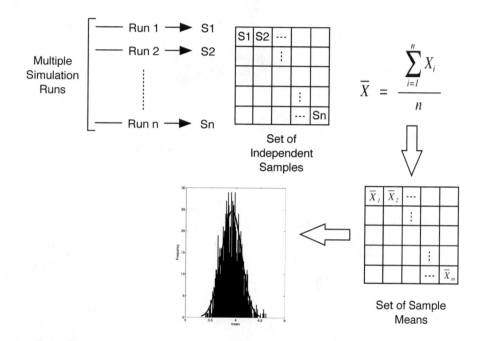

Figure 6.2
Probability distribution of the sample mean is normal.

6.2 PROBABILITY DISTRIBUTION OF THE SAMPLE MEAN

Suppose that n independent simulation runs of a simulation model are performed. As shown in Figure 6.2, at the end of every simulation run, a sample (S_i) of the performance measure of interest is generated. After all the n simulation runs are finished, we have a set of n samples. These samples constitute a sample set. If the above process is repeated m times, we end up with a set of sample means of size m. The frequency histogram of this set will have the shape of the normal distribution. This is the essence of the central limit theorem (see the next side note).

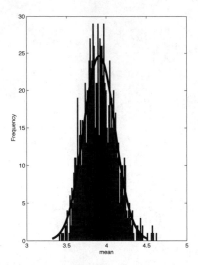

Figure 6.3
Frequency distribution of the average delay $\overline{D}$ through the single-server queue-ing system with $\lambda = 1$ and $\mu = 1.25$. The population mean is 4.

Figure 6.3 shows the frequency distribution of the average delay $(\overline{D})$ for the single-server queueing system. For this specific example, the population mean is 4. The population mean is equivalent to the theoretical mean which can be calculated using queueing theory. The standard deviation of this probability distribution is the standard error.

Now, since we know the probability distribution for $\overline{D}$, we can study how far the sample mean might be from the population mean. According to the empirical rule, approximately 68% of the samples fall within one standard deviation of the population mean. In addition, approximately 95% of the samples fall within two standard deviations of the population mean and approximately 99% fall within three standard deviations. Figure 6.4 illustrates the empirical rule. In the next section, we are going to use the fact that 95% of the samples lie within two standard deviations (i.e., $t = 1.96$) of the mean to establish a 95% confidence interval.

The Central Limit Theorem

Regardless of the probability distribution of the population mean, the probability distribution of the sample mean is always normal. The mean of this normal distribution is the theoretical mean and the standard deviation is the standard error.

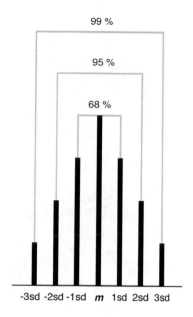

Figure 6.4
The empirical rule for the distribution of samples around the population mean.
95% of the area under the curve of the normal distribution lies within two
standard deviations (equal to 1.96) of the mean.

6.3 CONFIDENCE INTERVALS

A $((1 - \alpha) \times 100)\%$ confidence interval for a population mean μ is given by

$$\bar{x} \pm t \times \frac{s}{\sqrt{n}}, \tag{6.1}$$

where:

> t is a random variable that has a student-t distribution with $(n - 1)$
> degrees of freedom,
>
> $\bar{x}$ is the sample mean,
>
> s is the sample standard deviation,
>
> n is the sample size,
>
> α is the significance level, and
>
> $1 - \alpha$ is the confidence level.

From the above definition, it is clear that there are two factors that impact
the width of the confidence interval:

1. Confidence level $(1 - \alpha)$

   As the confidence level increases, the value of t increases. Accordingly, the width of the confidence interval increases.

2. Sample size (n)

   As the number of samples increases, the width of the confidence interval decreases.

Example 6.1 shows how different confidence intervals can be computed. Python has a statistics module that provides two methods for calculating the sample mean and standard deviation. Listing 6.1 shows how these two methods are used in the calculation of the confidence interval of a given sample set.

Example 6.1: Calculating the confidence interval.

Consider the following samples for estimating the average delay. Calculate the 80%, 90%, 95%, 98%, and 99% confidence intervals.

$$\{3.33, 3.15, 2.91, 3.05, 2.75\}$$

Solution

1. Calculate the sample mean and sample standard deviation.

$$\bar{x} = 3.038$$
$$s = 0.222$$

2. Compute the values of t for the different confidence levels. Using the t-distribution table in Appendix D and the fact that $n-1 = 4$, the values of t are as follows.

CL	t
0.80	1.533
0.90	2.132
0.95	2.776
0.98	3.747
0.99	4.604

Notice that as the confidence level increases, the value of t also increases.

3. Use Eqn. (6.1) to get the confidence intervals.

CL	t	Confidence Interval
0.80	1.533	(2.886, 3.190)
0.90	2.132	(2.826, 3.250)
0.95	2.776	(2.762, 3.314)
0.98	3.747	(2.666, 3.410)
0.99	4.604	(2.580, 3.495)

Notice that as the confidence level increases, the confidence interval gets wider.

Listing 6.1
Calculating the confidence interval using Python.

```python
import statistics as stat
import math


sample_set = [3.2, 3, 2.8, 2.9, 3.1]
n = len(sample_set)


mean = stat.mean(sample_set)
std_dev = stat.stdev(sample_set)


t = 2.776
ci1 = mean - t * (std_dev/math.sqrt(n))
ci2 = mean + t * (std_dev/math.sqrt(n))


print("Confidence Interval: ", round(ci1, 2), round(ci2, 2))


# Output
# Confidence Interval:  2.8 3.2
```

Note that when the number of samples is large (i.e., $n > 30$), the t-distribution approaches the normal distribution. As a result, the values of t become fixed. This is clearly shown in the last row of the table given in Appendix D.

6.3.1 Interpretations

The confidence interval is a random interval which may contain the population mean. The following is the mathematical expression for the probability that a confidence interval contains the population mean.

$$P[\bar{x} - t \times \frac{s}{\sqrt{n}} < \mu < \bar{x} + t \times \frac{s}{\sqrt{n}}] = 1 - \alpha,$$

where $\bar{x}$ is a random variable, μ is a constant, and $1 - \alpha$ is a probability. This expression says that the probability that the interval $(\bar{x} - t \times \frac{s}{\sqrt{n}}, \bar{x} + t \times \frac{s}{\sqrt{n}})$ contains μ is $1 - \alpha$. Therefore, we are $((1 - \alpha) \times 100)\%$ confident that the population mean is between $\bar{x} - t \times \frac{s}{\sqrt{n}}$ and $\bar{x} + t \times \frac{s}{\sqrt{n}}$. However, there is

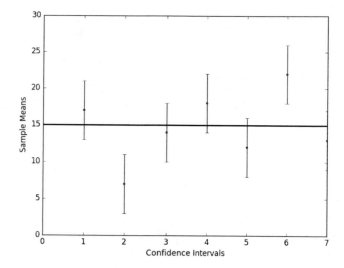

Figure 6.5
Two of the calculated confidence intervals do not include the population mean.
The population mean is 15.

a $(\alpha \times 100)\%$ chance that the population mean lies outside the confidence interval.

Another interpretation is the following. If a simulation is performed n times with different seed values, then in $((1-\alpha) \times 100)\%$ of the cases, the population mean lies within the confidence interval. In $(\alpha \times 100)\%$ of the cases, however, the population mean lies outside the interval. Figure 6.5 shows an example in which the confidence interval can miss the population mean. Listing 6.2 shows how this figure is generated.

Listing 6.2
Plotting confidence intervals and population mean.

```
1   import numpy as np
2   import matplotlib.pyplot as plt
3
4   x = [1, 2, 3, 4, 5, 6, 7]
5   y = [17, 7, 14, 18, 12, 22, 13]
6
7   plt.figure()
```

```
8   plt.plot([0, 7], [15, 15], 'k-', lw=2)   # Population Mean =
       15
9   plt.errorbar(x, y, yerr=4, fmt='.')
10  plt.xlabel("Confidence Intervals")
11  plt.ylabel("Sample Means")
12  plt.show()
```

The following are false interpretations of the confidence interval.

- The probability that the population mean belongs to the confidence interval is $(1 - \alpha)$.

  – This is wrong because the population mean is a constant that either belongs to the confidence interval or not.

- The percentage of the samples whose values are between $\bar{x} - t \times \frac{s}{\sqrt{n}}$ and $\bar{x} + t \times \frac{s}{\sqrt{n}}$ is $((1 - \alpha) \times 100)\%$.

  – This is wrong because the confidence interval is about the population, not the sample.

6.3.2   Why Not Always Use a 99% Confidence Interval?

First of all, we cannot have a 100% confidence interval. This is impossible because the population mean is unknown and we cannot generate all the members of a population. However, we can get the confidence level as high as 0.99. Unfortunately, as the confidence level increases, the confidence interval becomes wider and thus useless. For example, if you are 99% confident that the average delay is in the interval $(5, 90)$, then you cannot make any useful conclusion because the interval is simply too wide.

Example 6.2: Interpreting the confidence interval.

You have been asked to evaluate the performance of five machines. The 95% confidence interval for the average performance of one machine is (10.3, 13.1). Evaluate the following statements:

1. You are 95% confident that the performance for all the five machines is between 10.3 and 13.1.

2. 95% of all the samples generated as a result of running one machine will give an average performance between 10.3 and 13.1.

3. There is a 95% chance that the true average is between 10.3 and 13.1.

Solution

1. This is not correct: a confidence interval is for the population parameter, and in this case the mean, not for individuals.

2. This is not correct: each sample will give rise to a different confidence interval and 95% of these intervals will contain the true mean (i.e., the population mean).

3. This is not correct: μ is not random. The probability that it is between 10.3 and 13.1 is 0 or 1.

Example 6.3: Making a decision using the confidence interval.

A sales person working for a big manufacturer of networking devices claims that their new home router has an average delay of 54 msec. You decided to test the device in the lab. Based on a sample set of size 100, you have found that the average delay is 57 msec with a standard deviation of 1.2 msec. Would you buy this device?

Solution

1. Construct the 95% confidence interval for the mean delay.

$$(57 - 0.2352, 57 + 0.2352)$$

2. The confidence interval does not support the claim of the sales person because it does not contain the claimed population mean. Therefore, do not buy.

6.4 COMPARING TWO SYSTEM DESIGNS

Design means deciding the values of the system parameters (e.g., service rate) before the system is put into operation. Two system designs can be compared on the basis of a performance measure such as response time. However, you do not simply compare the mean values resulting from simulating the two designs and pick the best one. You need to first confirm whether the difference is caused by a difference in the design or by the random fluctuation inherent in the simulation model of the design. In this section, you are going to learn about how you select between two system designs using simulation.

Let the performance measure value obtained from simulating design i ($i \in \{1, 2\}$) be denoted by θ_i. Also, let the difference between the two performance measures be denoted by $\theta = \theta_1 - \theta_2$. We compute the confidence interval for θ to check if there is a significant difference between the two system designs. There will be three possible cases as follows.

1. If the confidence interval for θ contains zero, then the difference $\theta_1 - \theta_2$ is not statistically significant. Hence, there is no strong evidence that the observed difference is due to anything other than random variation in the output variables.

2. If the confidence interval for θ is to the left of zero, then there is a strong statistical evidence that $\theta_1 - \theta_2 < 0$. This means that the performance measure value for design 1 is smaller than that for design 2. Hence, design 1 is better.

3. If the confidence interval for θ is to the right of zero, then there is a strong statistical evidence that $\theta_1 - \theta_2 > 0$. This means that the performance measure value for design 2 is smaller than that for design 1. Hence, design 2 is better.

Example 6.4: Which design reduces the response time?

The table below shows five samples of the response time for two designs. In each simulation run, the same set of random numbers is used in simulating the two designs. The last column gives the difference in the response time of the two designs.

Replication	Design 1 (θ_1)	Design 2 (θ_2)	Difference ($\theta_1 - \theta_2$)
1	24	21	3
2	23	20	3
3	23	21	2
4	22	21	1
5	22	20	2
Mean			2.2
Variance			0.7
Standard Deviation			0.84

The 95% confidence interval for $\theta_1 - \theta_2$ is 2.2 ± 1.04, which is equivalent to $(1.16, 3.24)$. This implies that $\theta_1 - \theta_2 > 0$ and thus $\theta_1 > \theta_2$. Hence, design 2 is better and it will result in a smaller response time.

6.5 SUMMARY

Statistical analysis of the data resulting from running a simulation model is a very important step in a simulation study. This step will enable you to gain insights about the performance of the system you study. As a rule of thumb, for each performance measure you want to compute, you should report its mean and confidence interval. The confidence interval gives a range of values which may include the population mean. It enables you to assess how far your estimate (i.e., the sample mean) is from the true value (i.e., the population mean) of the performance measure.

6.6 EXERCISES

6.1 Simulate the single-server queueing system using 100 packets. Use $\lambda = 1$ and $\mu = 1.25$. Construct a 95% confidence interval for the average delay using a sample set of size 10. Answer the following questions:

a. Does the confidence interval include the theoretical mean?

b. If your answer in part (a) is NO, what do you think is the reason for missing the population mean? Suggest a solution.

II

Managing Complexity

## CHAPTER 7

Event Graphs

"All models are wrong, but some are useful."
−George Box

Event graphs are a formal modeling tool which can be used for building discrete-event simulation models. They were introduced in [6] as a graphical representation of the relationships between events in a system. Event graphs can be used to model almost any discrete-event system. This chapter is an introduction to event graphs. It is also going to show you how you can synthesize simulation programs from simulation models constructed using event graphs.

7.1 WHAT IS AN EVENT GRAPH?

An event graph is a visual representation of a discrete-event system. It shows the scheduling relationships between events which occur inside the system. An event graph is constructed using vertices and directed edges with attributes and conditions. In Figure 7.1(a), there are two events which are represented by two vertices A and B. Events A and B are referred to as the *source* and *target* events, respectively. When event A occurs, it will result in the scheduling of event B if the condition is true. Event B will occur after t time units.

On the other hand, in Figure 7.1(f), a dashed arrow is used to indicate that event A cancels event B. This type of edge is referred to as a *canceling* edge. A condition and time can be associated with a canceling edge. Similarly, for a normal edge, the condition and/or time can be eliminated. If both eliminated, that means the transition to the next event happens immediately (see Figure 7.1(d)).

Note that in Figure 7.1(b), event B is immediately scheduled if the condition on the scheduling edge evaluates to true. By contrast, in Figure 7.1(c), the scheduling edge is unconditional and event B will occur after t time units

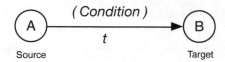

(a) Event B (target) is scheduled by event A (source) to occur after t time units if *condition* is true.

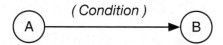

(b) Event B is going to occur immediately if *condition* is true.

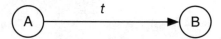

(c) Event B is going to occur after t time units.

(d) Event B is going to occur immediately.

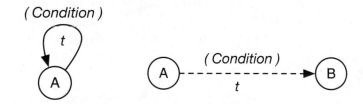

(e) Event A reschedules (f) A dashed arrow indicates that event B is
itself. cancelled after the occurrence of event A by t
time units if the *condition* is true.

Figure 7.1
Types of edges in event graphs.

of the occurrence of event A. Finally, an event can schedule itself as shown in Figure 7.1(e).

Event graphs also include state variables which are written between curly brackets. For example, in Figure 7.2(b), state variables are written below the vertices. The values the state variables should take are also shown. In this way, as we will see, we have a graph that describes the operation of the discrete-event system to be simulated. Once we are confident that the constructed event graph precisely captures the behavior of the system under study, we can call it a simulation model. Then, we translate the model into code and execute the simulation program.

Once the modeler has a good mental image of how the system under study works, he can proceed to construct a simulation model using event graphs. An event graph can be constructed as follows:

1. Identify all the event types in the system under study,

2. For each event type, identify the events it is going to schedule,

3. For each scheduling relationship between two events,

    (a) Identify the condition that must be satisfied in order for the target event to occur, and

    (b) Identify the time delay after which the target event occurs,

4. Each event type is represented by a vertex (i.e. node) in the event graph,

5. Vertices are connected by directed edges (either scheduling or canceling edges) that can have two attributes: delay and condition, and

6. Below each vertex, indicate the state variables which will be affected by the occurrence of the event and how they are updated.

7.2 EXAMPLES

In this section, several examples will be given to illustrate the modeling power of event graphs. These examples can be used as a basis for modeling more complex systems.

7.2.1 The Arrival Process

The arrival process is a very fundamental building block in event graphs of queueing systems. Its role is to generate packets and maintain a single state variable, A, which represents the cumulative number of packets. There is one random varaible, t_A, which is an exponential random variable with mean $\frac{1}{\lambda}$. It models the inter-arrival time between two consecutive arrivals.

Figures 7.2(a) and 7.2(b) show the state diagram and event graph of the Poisson arrival process, respectively. The event graph can be interpreted as

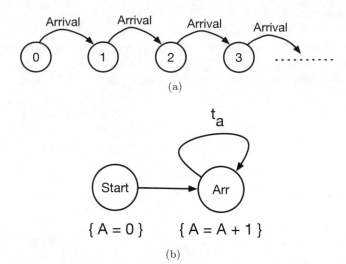

Figure 7.2
State diagram (a) and event graph (b) for the Poisson arrival process.

follows. First, when the *Start* event occurs at time $t = 0$, it sets the value of the state variable A to zero and schedules the first *Arrival* event to occur immediately at time $t = 0$. Then, whenever an arrival event occurs, the state variable A is incremented by one and the next arrival event is scheduled to occur after t_a time units.

7.2.2 Single-Server Queueing System

Figure 7.3 shows how the operation of the single-server queueing system is modeled as an event graph. There are four events:

1. Start,

2. Arrival (Arr),

3. Beginning of Service (Beg), and

4. End of Service or Departure (Dep).

There are two state variables: S and Q. The former represents the state of the server where $S = 0$ indicates that the server is *free* and $S = 1$ indicates that the server is *busy*. Once the state variables are initialized, the first *arrival* event is generated and the simulation is started. Whenever an arrival event occurs, the state of the server is checked. If the server is available, which is indicated by the condition $S == 0$, the service of the arriving packet is started immediately. Otherwise, the queue size is incremented by 1 (i.e., $Q = Q + 1$) to indicate that an arriving packet has been inserted into the queue. In

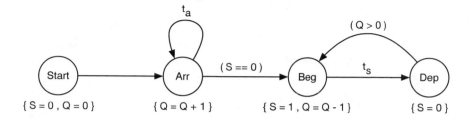

Figure 7.3
Event graph for the single-server queueing system.

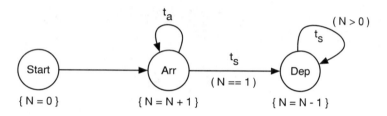

Figure 7.4
Reduced event graph for the single-server queueing system.

addition, the next arrival event is scheduled to occur after t_a time units. This is very important to keep the simulation running.

Next, whenever the service of a packet begins, the *Beg* event schedules a *Dep* event to occur after t_s time units. It also changes the state of the server to busy and decrements the size of the queue by one. Similarly, when the *Dep* event occurs, the state of the server is changed back to free and a new *Beg* event is scheduled if the queue is not empty.

The complexity of event graphs is measured by the number of vertices and edges present in the graph. Fortunately, an event graph can be reduced to a smaller graph, which is equivalent as far as the behavior of the system under study is concerned. However, it may be necessary to eliminate (and/or introduce new) state variables, attributes, and conditions in order to accommodate the new changes. In this specific example, the reduced event graph contains only one state variable, N, which represents the total number of packets inside the system.

Figure 7.4 shows a reduced event graph for the single-server queueing system. With this new event graph, the number of vertices and edges is reduced from four to three and five to four, respectively. Of course, for larger systems, the reduction will be significant.

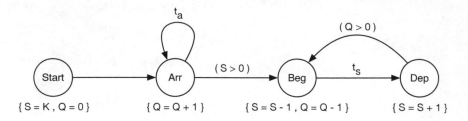

Figure 7.5
Event graph for the K-server queueing system.

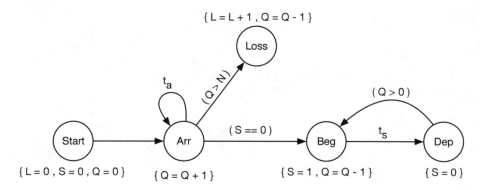

Figure 7.6
Event graph for the single-server queueing system with a limited queue capacity.

7.2.3 Multiple-Server Queueing System

A multiple-server queueing system is an extension of the single-server queueing system. It contains more than one server. Figure 7.5 shows the event graph for a multiple-server queueing system with K servers. The initial value of the state variable S is equal to the number of servers (K). An arriving packet will be scheduled for service as long as there is at least one available server. This is indicated by the condition $S > 0$. The value of S is decremented by one whenever the *Beg* event occurs. This is to indicate that a server has just been occupied. The value of S is incremented by one when a *Dep* event occurs to indicate that a server has just been released.

7.2.4 Single-Server Queueing System with a Limited Queue Capacity

The event graph in Figure 7.6 is for the single-server queueing system when the size of the queue is finite. In this kind of system, an arriving packet cannot be stored in the queue if it is full. As a result, the packet is lost. This requires

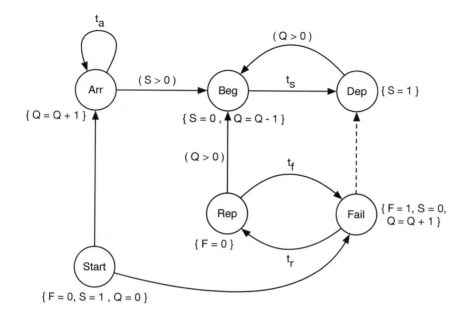

Figure 7.7
Event graph for the single-server queueing system with a server that fails.

the introduction of a new event type to capture this situation. This new event is referred to as the *Loss* event.

The *Loss* event occurs whenever there is an *Arrival* event and the number of packets in the buffer is N, which is the maximum queue size. When the arrival event occurs, the state variable Q is incremented by one and the next events are scheduled. If $Q > N$, a loss event is scheduled to occur immediately. When the loss event occurs, the state variable L is incremented by one to indicate a loss of a packet. On the other hand, the state variable Q is decremented by one since it has been incremented by one when the arrival event occurred.

7.2.5   Single-Server Queuing System with Failure

Consider a machine which fails periodically. When the machine fails, the part being manufactured is put back in the queue until the machine is repaired. Figure 7.7 shows the event graph for such a system. There are two initial events: *Arrival* and *Fail*. The failure event has the highest priority. That is, it will be executed before all events occurring at the same time.

When a fail event occurs, the current *Departure* event is canceled and a *Repair* event is scheduled instead. When the server becomes alive again, a *Beg* event for the part at the head of the queue is scheduled immediately. Also, the next failure event is scheduled to occur after t_f time units.

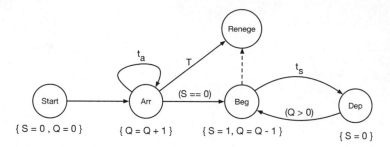

Figure 7.8
Event graph for the single-server queueing system with reneging.

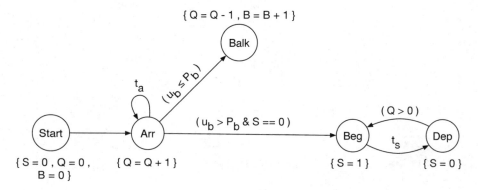

Figure 7.9
Event graph for the single-server queueing system with balking.

7.2.6 Single-Server Queuing System with Reneging

Impatient customers may leave the system after joining it. This can happen if a customer is not willing to wait for more than T time units. A customer *reneges* if he leaves the system without receiving service. Figure 7.8 shows the event graph for the single-server queueing system with reneging.

In this model, the renege time (i.e., T) is the same for all customers. This implies that the order of reneges is the same as the order of arrivals into the system. Hence, the canceling edge cancels the *Renege* event with the earliest simulation time. The result of a renege event is that the customer at the head of the queue is removed from the system.

7.2.7 Single-Server Queuing System with Balking

If the server in the single-server queueing system fails, customers may become unwilling to join the system. This is because their delay times will be severely affected. When a customer decides not to join the system, we say that the customer balks. That is, he refuses to enter the system and thus he leaves. So, balking means that the customer leaves the system upon arrival. A customer may balk with probability P_b or enter the system with probability $1 - P_b$. After he enters the system, a customer is either scheduled for service or he waits in the queue if the server is busy. Figure 7.9 shows the event graph for the single-server queueing system with balking. The state variable B is used to keep track of the customers who balk.

7.3   TRANSLATING EVENT GRAPHS INTO CODE

In this section, we are going to propose a set of high-level concepts which are expanded into code during the process of translating event graphs into idiomatic Python code. A piece of code is referred to as idiomatic if it represents what an experienced programmer would write. The proposed high-level concepts will help in mechanizing the translation process and enhancing the maintainability of the resulting code. These high-level concepts are the following:

1. Event type,

2. Event generator,

3. Event handler,

4. Initial event, and

5. Simulation loop.

An event type is a base concept and it includes two subconcepts: event generator and event handler. The event generator is an abstraction of a block of code which returns a realization (or instance) of an event type. This instance contains the time of occurrence of the event and the name of the event handler. An event type can be realized as a *tuple* in Python.

On the other hand, the event handler is an abstraction of a block of code which updates the state of the simulation model in response to an event. Two tasks are performed inside an event handler: (1) updating state variables and (2) scheduling next events. After an event handler is fully executed, control is returned to the main simulation loop.

Inside the simulation loop, the next event is always the one with the earliest occurrence time. If such an event exists, the simulation clock is updated by setting its value to the current simulation time, which is the time of occurrence of the current event. After that, the event handler of the event is called. This process is repeated until there are no more events to execute.

The notion of initial events is very crucial. These are the events which are placed inside the event list before the simulation loop is executed. They have to be explicitly identified in the event graph. In our case, we use a special event type called *Start* which points to the initial events. The start event always occurs at simulation time zero. When it occurs, it schedules the initial

Table 7.1
Event table for the event graph in Figure 7.4.

Event Type	Event Generator	Event Handler	State Variables
Start	NA	Handle_Start_Ev	$N = 0$
Arrival	Gen_Arr_Ev	Handle_Arr_Ev	$N = N + 1$
Departure	Gen_Dep_Ev	Handle_Dep_Ev	$N = N - 1$

events and starts the simulation. A block of code must exist in the simulation program before the simulation loop to explicitly place the initial events pointed to by the start event into the event list.

Listing 7.1
Python implementation of the event graph in Figure 7.4.

```python
from random import *
from bisect import *

# Parameters
lamda = 0.5
mu = 0.7
n = 100 # Number of packets to be simulated

# Initialization
clock = 0.0 # Simulation clock
evList = [] # Event list
count = 0    # Count number of packets simulated so far

# Insert an event into the event list
def insert(ev):
    insort_right(evList, ev)

# Event generator for the arrival event
def Gen_Arr_Ev (clock):
    global count
    count += 1
```

```
22        if count <= n:
23            ev = ( clock + expovariate(lamda) , Handle_Arr_Ev )
24            insert(ev)
25
26   # Event generator for the departure event
27   def Gen_Dep_Ev (clock):
28        ev = ( clock + expovariate(mu) , Handle_Dep_Ev )
29        insert(ev)
30
31   # Event handler for the arrival event
32   def Handle_Arr_Ev(clock):
33        global N
34        N = N + 1     # Update state variable
35        Gen_Arr_Ev(clock)    # Generate next arrival event
36        if N == 1:
37            Gen_Dep_Ev(clock)
38
39   # Event handler for the departure event
40   def Handle_Dep_Ev(clock):
41        global N
42        N = N - 1
43        if N > 0:
44            Gen_Dep_Ev (clock)
45
46   # Initialize state variables and generate initial events
47   N = 0                    # State variable
48   Gen_Arr_Ev(0.0)     # Initial event
49
50   # Simulation loop
51   while evList:
52        ev = evList.pop(0)
53        clock = ev[0]
54        ev[1](clock) # Handle event
```

Table 7.1 is the event table for the event graph in Figure 7.4. Listing 7.1 shows how the information in Table 7.1 is translated into Python code. As you can tell, the code is very structured. Thus, this process can be automated very easily. Next, this translation process is described.

In the the first part of the program (lines 1-2), standard Python libraries (*random* and *bisect*) are imported into the program. The first one contains functions which can be used for random number generation. The second one, however, contains functions for manipulating the event list. Then, parameters of the simulation models are defined on lines 5-7. There are only three parameters: arrival rate (*lamda*), service rate (*mu*), and number of packets to be simulated (*n*).

In the third part of the program (lines 10-13), the simulator is initialized. First, the simulation clock is set to zero. This variable is used to keep track of the simulation time. After that, an empty list is created to keep the simulation events. In this list, events are kept in order using the predefined function *insort'right* from the *bisect* library (see line 16). There is only one state variable, N, in this simulation model. It is initialized on line 47. The variable *count* defined on line 13 is used for counting the number of packets which have been simulated. This is necessary to make sure that no more than n packets are simulated. The value of this variable is incremented and checked inside the event generator of the arrival event (see lines 20-24).

In the fourth part of the program (lines 15-16), a convenience function is defined. This abstraction hides the unconventional name used for the predefined function used for sorting events in the event list. This part is not necessary. However, we believe it helps in enhancing the readability of the code.

Figure 7.10 shows a template which can be used as an aid when performing a manual translation of an event graph into a Python simulation program. The match between program in Listing 7.1 and the proposed template is almost perfect. Each block in the template corresponds to a modeling concept. This concept is expanded into code in the final simulation program.

7.4 SUMMARY

A visual simulation model of any discrete-event system can be constructed using event graphs. Although an event graph gives a very high-level representation of the system, it still helps in capturing and understanding the complex relationships between events inside the simulation model. In addition, event graphs can be translated into Python code in a systematic way using the abstractions discussed in this chapter. The resulting code is easy to understand and maintain. Of course, as the size of the system grows, its event graphs become very complicated.

7.5 EXERCISES

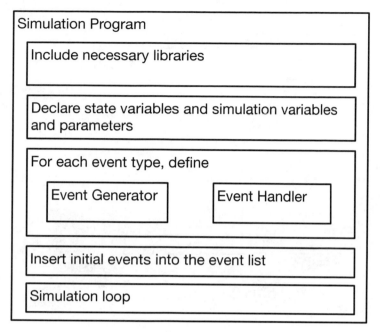

Figure 7.10
A template for synthesizing simulation programs from event graphs.

7.1 Consider the system in Figure 7.11. There are two independent single-server queueing systems. There is one traffic source which feeds the two systems. Traffic is randomly split between the two systems. That is, an arriving packet joins the system i with probability $\frac{\lambda_i}{\lambda}$, where $\lambda_1 + \lambda_2 = \lambda$.

a. Draw the event graph for the traffic generator.

b. Now, draw the event graph for the whole system.

7.2 Consider the setup in Figure 7.12 where a user communicates with an online service hosted in a data center. The channel connecting the user to the service has two characteristics: (1) propagation delay (P_d) and (2) rate (R). The propagation delay is the time required for an electrical signal to travel from the input of the channel to its output. Hence, if a packet is injected into the channel, it will arrive at the other end after P_d time units. The rate, however, is the speed of the channel. It gives the number of bits which can be injected into the channel in one second. Thus, $\frac{1}{R}$ is the time required to inject (i.e., transmit) one bit. The user communicates with the server using a simple protocol. Basically, the user transmits a message. Then, it waits until it receives an acknowledgment from the server that the message has been received successfully. If the user does not receive an acknowledgment within a preset time period,

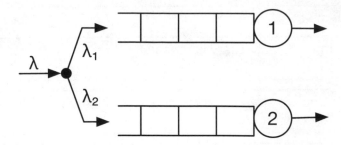

Figure 7.11
Two parallel single-server queueing systems with one shared traffic source.

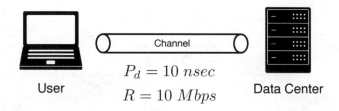

Figure 7.12
A simple network setup where a user communicates with a server in a data center over a communication channel created inside a network. Propagation delay (P_d) and rate (R) are two important characteristics of a channel.

it retransmits the same message again. Otherwise, it transmits the next message.

a. Identify all the possible events which can occur in this system.

b. Draw an event graph which describes the operation of this system.

7.3 Consider the event graph for the single-server queuing system with reneging (see Figure 7.8). Assume that the renege time of each packet is different. In this case, the order of reneges is not necessarily the same as the order of arrivals to the system. Modify the event graph in Figure 7.8 to reflect this new situation.

## CHAPTER 8

Building Simulation Programs

"Messy code often hides bugs."
—Bjarne Stroustrup

Simulation programs are either time-driven or event-driven. In both cases, the state of the simulation model is updated at discrete points in time. In this chapter, you will learn the difference between the two types of programs. The emphasis, however, will be on event-driven simulation. The general structure of any discrete-event simulation program will be discussed. A complete program will be shown and several programming issues will be pointed out.

8.1 TIME-DRIVEN SIMULATION

This approach to simulation is also referred to as *discrete-time* simulation. This is because time evolves in discrete steps as shown in Figure 8.1. In this case, the time axis is divided into equal intervals called *slots*. The size of each slot is Δt. Events occur at the boundaries of each slot (either at the beginning or end of the slot). For example, the arrival of a packet can occur at the beginning of a slot while the departure occurs at the end of the slot. State variables, on the other hand, are updated at the end of the slot after all events have occurred.

Slots are sequentially numbered using non-negative integers. Slot n is located in time $[n-1, n)$, where $n = 1, 2, 3, \ldots$. Hence, in a time-driven simulation program, the simulation loop has the following form:

```
for n in range(1, Total_Number_Of_Slots):
    clock = clock + Delta_T
```

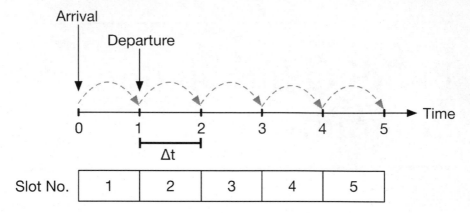

Figure 8.1
In time-driven simulation, simulated time evolves in increments of size Δt.

The variable `clock` represents the current simulated time, which advances in increments of size `Delta_T`.

Listing 8.1 shows a time-driven simulation program for a discrete-time model of the single-server queueing system. In this case, the arrival and departure processes are Bernoulli random processes (see Figure 4.14). In each time slot, an arrival and departure can occur with probabilities P_a and P_d, respectively. The system will be simulated for a period of `Total_Number_Of_Slots` slots. There is only one state variable which is Q.

The simulation loop starts on line 15. In each iteration, the simulated time is updated by `Delta_T`. Then, a random number is generated to check if an arrival has occurred (see line 17). The auxiliary variable `A` is set to one if there is an arrival. Similarly, on lines 21 and 22, a random number is generated and the auxiliary variable `D` is set to one if there is a departure. Finally, the state variable `Q` is updated at the end of the simulation loop.

Listing 8.1
A time-driven simulation program for the discrete-time single-server queueing system.

```
1  from random import *
2  from statistics import *
3
4  Pa = 0.2 # Probability of an arrival
5  Pd = 0.6 # Probability of a departure
6
```

```
7   clock = 0
8   Delta_T = 1
9
10  Total_Number_Of_Slots = 10000
11
12  Q = 0 # Number of packets in queue
13
14  # Simulation loop
15  for n in range(1, Total_Number_Of_Slots):
16      A = 0 # Auxiliary variable for indicating an arrival
17      D = 0 # Departure
18      clock = clock + Delta_T
19      if random() <= Pa:
20          A = 1
21      if random() <= Pd and Q > 0:
22          D = 1
23      # Update state variable
24      Q = Q + (A - D)
```

Continuous-Time Versus Discrete-Time Queues

The single-server queueing system can be modeled in two ways: discrete-time and continuous-time. In discrete-time queues, time evolves in discrete steps of the same size (see Figure 8.1). On the other hand, in continuous-time queues, events can occur at any point on the time line. Hence, the time between two consecutive events is random (see Figure 8.3). In addition, the arrival and departure processes along with their underlying random variables are different (see Figure 8.2) .

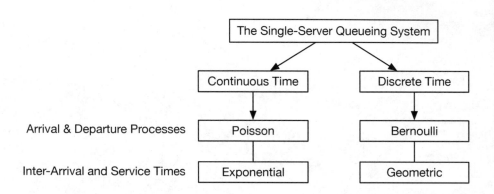

Figure 8.2
Arrival and departure processes and their random variables in continuous- and discrete-time queues.

8.2   EVENT-DRIVEN SIMULATION

This approach to simulation is also called *discrete-event* simulation. In this type of simulation, time evolves in discrete steps of random sizes. Hence, as shown in Figure 8.3, events occur at random points along the time line. Also, the time between two consecutive events is random. The state variables must be updated after the occurrence of every event.

In an event-driven simulation program, the simulation loop has the following form:

```
while Event_List NOT Empty:
    ev = EventList.next()
    clock = ev.time
```

Clearly, the simulated time will advance in steps of unpredictable (i.e., random) sizes. Also note that a WHILE loop is used instead of a FOR loop. This loop is terminated when the list of events become empty.

Every event-driven simulation program must contain an event list. This list maintains the temporal order of events. The first event in the list is always the

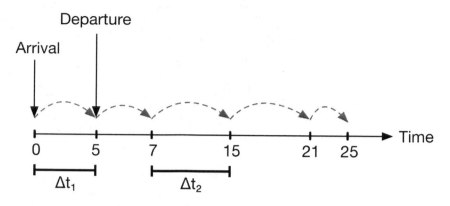

Figure 8.3
In event-driven simulation, simulated time evolves in steps of random sizes $(\Delta t_1 \neq \Delta t_2)$.

one with the earliest event time. Fortunately, in Python, this list is already implemented for us. For more details, see section A.7 in Appendix A.

8.3   WRITING EVENT-DRIVEN SIMULATION PROGRAMS

Now, we formally define the structure of an event-driven simulation program. As shown in Figure 8.4, an event-driven simulation program consists of two components: simulator and model. Events are generated by the model. They are applied back to the model by the simulator. The simulator is responsible for maintaining the temporal order of events using the event list. It is also responsible for keeping the current simulated time (also called simulation time) uptodate. At the beginning, the event list will contain a set of initial events which will be used to start up the simulation.

The simulator contains a Random Number Generator (RNG), which is the main source of randomness in the simulation program. The role of the RNG is to generate random numbers from the interval $(0, 1)$. These random numbers are then used to drive the Random Variate Generators (RVG) in the simulation model. RVGs and RNGs will be discussed in Chapters 10 and 11, respectively. Figure 8.5 illustrates the relationships between RNG, RVG, and Random Event Generator (REG). REGs are responsible for generating events in the simulation program. For each event type, there will be a separate REG.

The model is the conceptual representation of the system being simulated. Its elements are specific to the system and they must capture its behavior. The model is executed by applying events to it. This will result in changing the values of the state variables in the model. State variables are updated inside blocks of code referred to as event handlers. There will be one event handler for each event type.

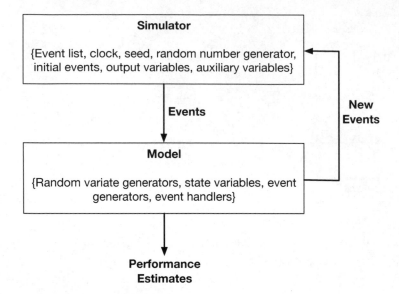

Figure 8.4
An event-driven simulation program has two independent components: simulator and model.

Figure 8.5
How a random number u is used to generate an event.

Executing the simulation model results in new events which are passed to the simulator. After sorting them, the simulator applies them back to the simulation model. Whenever an event is applied, the current values of some state variables are recorded in predefined output variables. These values (or samples) will eventually be used for computing statistics about the performance of the system under study.

Figure 8.6 shows the general structure of any event-driven simulation program. The two components mentioned above are explicitly identified. There are mainly four steps. Step 1 and 2 are part of the simulator. In Step 1, the program is initialized. Parameters are read from the user and variables are declared. The event list is created and initial events are inserted into it. Finally, the simulation clock is set to zero. After that, in Step 2, the simulation loop is executed. In each iteration of this loop, the next event is fetched from the event list. It is the event with the earliest event time. The clock is updated and the event handler of the event is called.

In Step 3, the model is executed as a result of calling event handlers. Inside

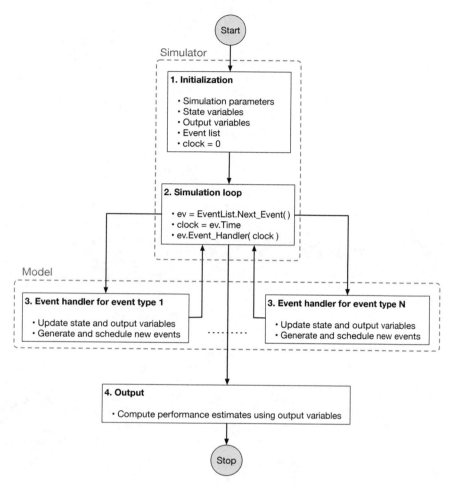

Figure 8.6
A flowchart of the event-driven simulation program.

each event handler, state variables are first updated. Then, they are sampled and their current values are stored in the corresponding output variables. Finally, new events are generated and they are passed to the simulator. Steps 2 and 3 are executed repetitively until the condition of the simulation loop becomes true. For instance, the simulation loop should be terminated when the event list becomes empty. In the final step of the program (i.e., Step 4), statistical estimates of the performance measures are computed using the values of state variables stored in the output variables.

Table 8.1

Mapping concepts to code in Listing 8.2.

Operations	Lines
Initialization	6 - 23
REG for the arrival event	25 - 29
REG for the departure event	31 - 35
Event handler for the arrival event	37 - 47
Event handler for the departure event	49 - 58
Insert initial events into the event list	61 - 63
Simulation loop	80 - 83
Statistical summaries	96 - 104

Listing 8.2

An event-driven simulation program for the single-server queueing system.

```python
1  from random import *
2  from queue import *
3  from statistics import *
4  from math import *
5
6  # Simulation parameters
7  lamda = 0.2
8  mu = 0.3
9  n = 10000   # Number of simulated packets
10
11 # Unique ID for each event
12 evID = 0
13
14 # Count number of simulated packets
15 count = 0
16
17 # State variables
18 Q = 0
19 S = False   # Server is free
20
21 # Output variables
22 arrs = []
```

```python
23   deps = []
24
25   # Event list
26   evList = None
27
28   # REG for the arrival event
29   def get_next_arrival_event (clock):
30       global evID
31       iat = expovariate(lamda)
32       ev = ( clock + iat , evID, arrival_event_handler )
33       evID += 1
34       return ev
35
36   # REG for the departure event
37   def get_next_departure_event (clock):
38       global evID
39       st = expovariate(mu)
40       ev = ( clock + st , evID, departure_event_handler )
41       evID += 1
42       return ev
43
44   # Event handler for the arrival event
45   def arrival_event_handler (clock):
46       global n, count, Q, S, arrs
47       Q += 1
48       arrs.append(clock)   # Record arrival time
49       if S == False:
50           S = True
51           schedule_event ( get_next_departure_event(clock) )
52       count += 1
53       if count < n:
54           schedule_event ( get_next_arrival_event(clock) )
55
56   # Event handler for the departure event
```

```python
57  def departure_event_handler (clock):
58      global Q, S, deps
59      Q -= 1
60      deps.append(clock)   # Record departure time
61      if Q == 0:
62          S = False
63      else:
64          S = True
65          schedule_event( get_next_departure_event(clock) )
66
67  # Insert an event into the event list
68  def schedule_event(ev):
69      global evList
70      evList.put(ev)
71
72  # Main simulation function
73  def sim():
74      global Q, S, arrs, deps, count, evList
75      clock = 0
76      evList = PriorityQueue()
77      # Reset state and output variables
78      Q = 0
79      S = False
80      arrs = []
81      deps = []
82      count = 0
83      # Insert initial events
84      ev = get_next_arrival_event(clock)
85      schedule_event(ev)
86      # Start simulation
87      while not evList.empty():
88          ev = evList.get()
89          clock = ev[0]
90          ev[2](clock)
```

```python
def main():
    global arrs, deps
    m = 50  # Number of replications
    Samples = []
    for i in range(m):
        d = []
        seed()  # Reseed RNG
        sim()
        d = list( map(lambda x,y: x-y, deps, arrs) )
        Samples.append( mean(d) )

    sample_mean = mean(Samples)
    sample_std_dev = stdev(Samples)
    t = 1.96
    ci1 = sample_mean - t * (sample_std_dev / sqrt(m))
    ci2 = sample_mean + t * (sample_std_dev / sqrt(m))

    print( "Average Delay = ", round(sample_mean, 2) )
    print( "Confidence Interval: ", "( ", round(ci1, 2), ",
    ", round(ci2, 2), " )" )
    print( "Population Mean = ", round(1 / (mu-lamda), 2) )

if __name__ == '__main__':
    main()

### Example output
# Average Delay =  10.09
# Confidence Interval:  (  9.96 ,  10.21  )
# Population Mean = 10.0
```

8.4 PROGRAMMING ISSUES

Several programming issues arise when writing event-driven simulation programs. The following are three critical issues that must be handled appropriately. Mishandling them may cause the simulation program to produce wrong statistical results.

8.4.1 Event Collision

An event is represented by a tuple inside each event generator (e.g., see lines 28 and 34). When inserted into the event list, the first field in the tuple is used as a key for sorting the event. When two events have the same key, an *event collision* is said to have occurred. Thus, the next field in the tuple is used as a key. By convention, the first field in the tuple representing an event is the time of the event. The second field is an event identifier. This is a unique key which maintains the order in which events are generated. In this way, it is guaranteed that no event collision will occur.

8.4.2 Identifiers for Packets

When recording data in output variables, the order of packets must be maintained. That is, the ith entry in any output variable must correspond to the ith simulated packet. If this order is not maintained, the final statistical results will be wrong. In the case of the single-server queueing system, maintaining the order is easy. This is because packets leave in the same order in which they enter the system. Hence, on lines 42 and 53 in Listing 8.2, event times are appended to the end of output variables. The index of each entry in each list corresponding to an output variable represents the identifier of the packet.

8.4.3 Stopping Conditions for the Simulation Loop

There are several options to terminate a simulation loop. A simulation loop can be terminated when the

1. Event list becomes empty,

2. Number of simulated packet reaches a preset value, and

3. Maximum allowed simulation time is reached.

In Listing 8.2, the simulation loop is terminated when the event list becomes empty (see line 80). In order to allow this, there should be a limit on the number of packets to be simulated (i.e., n). The variable count is used to keep track of the number of arrivals. This variable is incremented inside the event handler of the arrival event (see lines 52-53). No more arrivals will be generated if the number of arrivals of n is exceeded. This will guarantee that the simulation program will terminate once the event list becomes empty. In addition, it is guaranteed that all the generated n packets will be simulated.

If there is a limit on the number of simulated packets, the condition of the simulation loop will become as follows:

```
Total_Num_Pkts = 1000
while Num_Simulated_Pkts <= Total_Num_Pkts:
    ...
```

The variable Num_Simulated_Pkts must be incremented inside the event handler of the departure event (i.e., the function departure_event_handler). This is because this function represents the exit point for each packet from the system. For each packet, when the part of the simulation program is reached, it means that the life cycle of the packet has been fully simulated.

On the other hand, if the limit is on the total simulation time, the condition of the simulation loop will become as follows:

```
Total_Sim_Time = 1000
while clock <= Total_Sim_Time:
    clock = clock + ev.time
```

Note that there might be some packets pending in the event list. Therefore, you will have to keep track of the number of packets which have left the system. This number represents the number of entries in the output variables which you can use in computing the statistical results.

8.5 SUMMARY

There are two approaches to writing simulation programs: time-driven and event-driven. The second approach is the most common one. A template for discrete-event simulation programs was proposed in this chapter. In addition, several programming issues were mentioned and their solutions were suggested.

8.6 EXERCISES

8.1 Write a time-driven simulation program for the single-server queueing system where the arrival process is Poisson. Assume that in each time slot, one departure will occur if the queue is not empty. Compute the average delay through this system.

8.2 Consider the system configuration in Figure 8.7. Write a discrete-event simulation program that simulates this system and computes the average delay through it.

8.3 Consider the single-server queueing system with reneging (see Figure 7.8). After waiting for five minutes in the queue, a customer reneges. Write a discrete-event simulation program to estimate the average time between customers who renege.

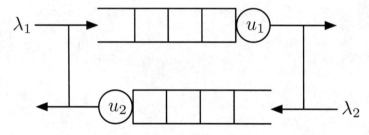

Figure 8.7
Two single-server queueing systems in series with external arrivals.

III

Problem-Solving

## CHAPTER 9

The Monte Carlo Method

"There is no such thing as luck. It is all mathematics."
−Nico Zographos

The Monte Carlo (MC) method was born during the second world war. It was used in the simulation of atomic collisions which then resulted in the first atomic bomb. Nowadays, the MC method is used in different fields such as mathematics, physics, biology, and engineering. In its simplest form, *a MC method is an algorithm that use random variates to compute its output.* In this chapter, we are going to explore through concrete applications the usefulness of the MC method. In addition, several enhanced versions of the original MC method are discussed.

9.1   ESTIMATING THE VALUE OF π

The MC method can be used to estimate the value of a parameter or constant. In this section, you are going to learn how the setup shown in Figure 9.1 can be used to estimate the value of π, which is the ratio of the circumference of a circle to its diameter. π is approximately equal to *3.14*.

Consider a circle with radius r and centered at the origin as shown in Figure 9.1. This circle is also enclosed inside a square with an edge length of $2r$. A point (x, y) falls inside the circle if the following inequality is satisfied:

$$x^2 + y^2 \leq r^2 \tag{9.1}$$

Both x and y take values from the interval $[-1, +1]$. r has a fixed value of 1.

In Figure 9.1, there are two regions: Circle (C) and Square (S). S contains C. From measure theory, the probability that a point (x, y) lies inside C is

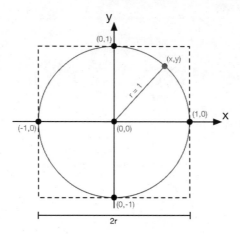

Figure 9.1
Setup used for performing MC simulation to estimate π.

given by:

$$
\begin{aligned}
P[(x,y) \in C] &= \frac{measure\ of\ C}{measure\ of\ S} \\
&= \frac{area\ of\ C}{area\ of\ S} \\
&= \frac{\pi r^2}{4r^2} \\
&= \frac{\pi}{4}.
\end{aligned}
\tag{9.2}
$$

Hence, the following equation for π can be deduced.

$$
\pi = 4 \cdot P.
\tag{9.3}
$$

Now, we have an expression for π. However, we still need to estimate the value of P. Since P is the probability of an event, a binary (i.e., Bernoulli) random variable should be used in the simulation. This variable is defined as follows:

$$
Z = \begin{cases} 1, & \text{if } (x,y) \in C, \\ 0, & \text{otherwise.} \end{cases}
\tag{9.4}
$$

The expected (i.e., average) value of Z represents the value of P. It is the proportion of times the event of interest (i.e., $\{(x,y) \in C\}$) occurs in a long series of trials. It can mathematically be expressed as follows:

$$
\begin{aligned}
E[Z] &= 1 \cdot P[\{(x,y) \in C\}] + 0 \cdot P[\{(x,y) \notin C\}] \\
&= P[\{(x,y) \in C\}] \\
&= \frac{\pi}{4}.
\end{aligned}
\tag{9.5}
$$

> **Listing 9.1**
> Python procedure for estimating π using MC simulation.

```python
from random import *
from statistics import *


N = 100000


Z = []
for i in range(N):
    x = uniform(-1, 1)
    y = uniform(-1, 1)
    if x**2 + y**2 <= 1:
        Z.append(1)
    else:
        Z.append(0)

print ("Pi = ", 4.0 * round(mean(Z), 4))      # = 3.1452
print ("Variance = ", round(variance(Z), 4))  # = 0.1681
```

Hence, π can be estimated using the following estimator:

$$\pi = 4 \cdot E[Z]. \tag{9.6}$$

$E[Z]$ can be approximated using Monte Carlo simulation as follows:

$$E[Z] \approx \frac{1}{N} \sum_{i=1}^{N} Z_i \tag{9.7}$$

where Z_i is a Bernoulli random variate generated using Eqn. (9.4).

Listing 9.1 shows a Python procedure that approximates the value of π using Monte Carlo simulation. This version of Monte Carlo simulation is referred to as the *Crude* Monte Carlo (CMC) simulation. It is crude because it typically results in a high variance (see line 16).

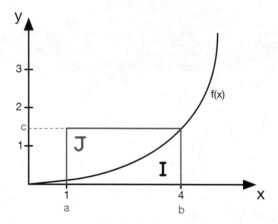

Figure 9.2
Setup used for performing MC simulation to estimate a one-dimensional integral.

9.2 NUMERICAL INTEGRATION

Figure 9.2 shows a function $f(x)$ defined over the interval $[a, b]$. The function $f(x)$ is also enclosed inside a rectangle with width $b - a$ and height c. The curve of $f(x)$ divides the rectangle into two regions I and J. Region I is the one under the curve. We want to find the area of this region. This area is mathematically defined as follows:

$$A_I = \int_a^b f(x) \ dx. \tag{9.8}$$

The probability that a randomly generated point falls inside region I can be computed as follows:

$$P[(x, y) \in I] = \frac{measure\ of\ region\ I}{measure\ of\ region\ J}$$
$$= \frac{area\ of\ region\ I}{area\ of\ region\ J} \tag{9.9}$$
$$P = \frac{A_I}{A_J},$$

where the area of region J is equal to $A_J = c \cdot (b - a)$.

Hence, the integral can be estimated using the following estimator:

$$A_I = P \cdot [(b - a) \cdot c] \tag{9.10}$$

$$P = E[Z]$$

$$\approx \frac{1}{N} \sum_{i=1}^{N} Z_i \tag{9.11}$$

where Z_i is a Bernoulli random variate that can be generated using the following equation:

$$Z = \begin{cases} 1, & \text{if } (x, y) \in I, \\ 0, & \text{otherwise.} \end{cases} \tag{9.12}$$

Listing 9.2 shows how a one-dimensional integral can be estimated using the CMC method.

Listing 9.2
Python procedure for estimating a one-dimensional integral.

```python
from random import *
from statistics import *

# Specify parameters
a = 1
b = 8
N = 100000

# Integrand
def f(x):
    return x**2

# Find value of c
c = f(b)

# Area of rectangle
A_J = (b-a) * c

Z = [0]*N
for i in range(N):
    x = uniform(a, b)
    y = uniform(0, c)
```

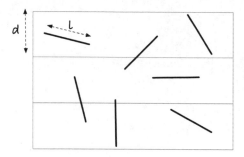

Figure 9.3
The goal of the Buffon's needle experiment is to compute the probability that
a needle of length l will intersect a horizontal line in a set of horizontal lines
separated by a distance equal to d.

```
23        if y <= f(x):
24            Z[i] = 1
25
26    A_I = mean(Z) * A_J
27
28    print("A_I = ", round(A_I, 2))   # =   169.57 (170.33)
29    print("Variance = ", round(variance(Z), 4)) # = 0.2352
```

9.3   ESTIMATING A PROBABILITY

The CMC method can also be used for estimating probabilities of events.
These events should occur with high frequency. An event that occurs with a
low frequency is referred to as a *rare* event. The CMC method usually fails
in estimating probabilities of rare events. Advanced MC techniques are used
instead, as will be shown in the next section.

9.3.1   Buffon's Needle Problem

In this problem, a needle of length l is dropped onto a floor with equally spaced
horizontal lines. That is, the distance between every two consecutive lines is d.
The length of the needle is constrained such that $l \le d$. The goal is to estimate
the probability that the needle touches or intersects a line. Figure 9.3 shows
the setup of the problem along with some needles at random locations.

The simulation model for this experiment makes use of two random vari-

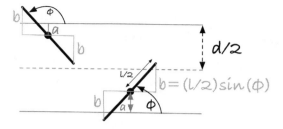

Figure 9.4
Two random variables (a and ϕ) are used in the simulation. The needle will intersect with the closest horizontal line if $b \geq a$.

ables. These two random variables uniquely identify the location of the needle on the floor. The two random variables are the following:

a: Distance from the midpoint of the needle to the closest horizontal line ($a \in [0, \frac{d}{2}]$)

θ: Angle the needle makes with the closest horizontal line ($\theta \in [0, \pi]$)

Figure 9.4 shows a portion of the floor with one needle and two horizontal lines. It also shows how the two random variables defined above are used to characterize the location of the needle. Clearly, the needle will intersect a horizontal line if $a \leq b$. Figure 9.5 is a reminder of how the value of b can be computed by using basic trigonometry. The exact expression for the probability is the following [5]:

$$P = \frac{2l}{\pi d}. \tag{9.13}$$

Listing 9.3
Python procedure for the Buffon's needle experiment.

```python
from random import *
from math import *
l = 1
d = 1
n = 1000000
count = 0
for i in range(n):
    a = uniform(0, d/2)
```

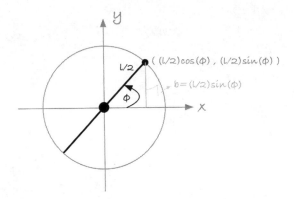

Figure 9.5
According to trigonometry, the length of the line segment b is equal to the value of the y-coordinate of the upper tip of the needle.

```
9    phi = uniform(0, pi)
10   b = (1/2)*sin(phi)
11   if a <= b:
12       count = count + 1
13   print('P = ', round(count/n, 3))
14   print('Exact = ', round((2*l)/(pi*d), 3))
```

9.3.2 Reliability

Consider the block in Figure 9.6(a) where the input is connected to the output if the switch is closed. The probability of this event (i.e., switch is closed) corresponds to the portion of time the block is working. Let R be the reliability of a block. Then, the reliability of the system (i.e., Rel_{sys}) in Figure 9.6(b) is R^3. It is the product of the reliability of the three blocks in series. Next, we develop a simulation model to computationally estimate this number. Since we know the exact answer in advance, we can easily tell if the proposed Python procedure is correct.

First, let us define the sample space of the problem. The state of the system (denoted by s_i) is a set of three random variables (denoted by b_1, b_2, and b_3), where each random variable corresponds to the state of an individual block in

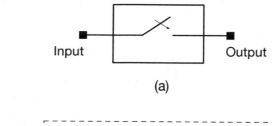

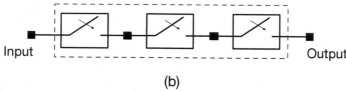

Figure 9.6
Reliability is the probability that the input is connected to the output. (a) The input is connected to the output if the swtich is closed. (b) Reliability of the overall system is a function of the reliabilities of the individual blocks. In this case, $Rel_{sys} = R^3$, where R is the reliability of a block.

the system. Each b_i is a binary random variable defined as follows:

$$b_i = \begin{cases} 1, & \text{with prob. } p \\ 0, & \text{with prob. } q = 1 - p. \end{cases} \tag{9.14}$$

Table 9.1 shows the individual points in the sample space, which is of size 2^3. It also shows the probability of each possible systems state.

Now, let us define a new random variable ϕ over the sample space of system states. This random variable is defined as follows:

$$\phi(s_i) = \begin{cases} 1, & \text{if input is connected to output} \\ 0, & \text{otherwise.} \end{cases} \tag{9.15}$$

Next, the system reliability can be calculated as follows:

$$Rel_{sys} = E[\phi]$$
$$= \sum_{i=1}^{2^3} \phi(s_i)P[s_i]. \tag{9.16}$$

The random variable ϕ will be one for s_8 only. This event occurs with a probability of $p^3 = 0.343$. Hence, the reliability of the system is calculated as follows:

Table 9.1
Sample space of the system in Figure 9.6(b) with probability of each sample point.

System State	b_1	b_2	b_3	$P[s_i]$
s_1	0	0	0	q^3
s_2	0	0	1	q^2p
s_3	0	1	0	q^2p
s_4	0	1	1	qp^2
s_5	1	0	0	pq^2
s_6	1	0	1	p^2q
s_7	1	1	0	p^2q
s_8	1	1	1	p^3
				$\sum P[s_i] = 1.0$
				$Rel_{sys} = 0.343$

$$Rel_{sys} = E[\phi]$$
$$= \sum_{i=1}^{7} \phi(s_i) \cdot P[s_i] + \phi(s_8) \cdot P[s_8]$$
$$= \sum_{i=1}^{7} 0 \cdot P[s_i] + 1 \cdot P[s_8]$$
$$= 1 \times 0.343.$$
$$= 0.343.$$

Listing 9.4 shows how this reliability can be estimated using the CMC method. A realization of the system is generated on lines 15-17. Then, the realization is checked if it represents a connected system, which is the event of interest.

Listing 9.4
Estimating the reliability of the system in Figure 9.6(b).

```
1   from random import *
2
3   Num_Trials = 100000
4   count = 0
5   p = 0.3    #Probability a block is working
6
7   def Phi(X):
```

```
 8      if sum(X) == 3:
 9          return 1
10      else:
11          return 0
12
13  for i in range(Num_Trials):
14      X = []
15      for j in range(3):
16          if random() <= p: X.append(1)
17          else: X.append(0)
18      count = count + Phi(X)
19
20  print('Rel_sys = ', round(count / Num_Trials, 3))
```

9.4   VARIANCE REDUCTION TECHNIQUES

The CMC method may require a very large number of samples in order to generate an acceptable result. In other cases, it may fail if the event of interest is rare. This is why advanced MC methods are needed. The advanced versions of the CMC method can achieve the same level of accuracy using a smaller number of samples. They can also be used for estimating probabilities of rare events by changing the probability distribution of the event of interest.

9.4.1   Control Variates

Consider a random variable X whose expected value $E[X]$ is to be estimated. Assume there is another random variable Y whose expected value $E[Y]$ is known. Then, the following is an estimator of $E[X]$:

$$E[X] = \frac{1}{n}\sum_{i=1}^{n} X_i - c\left(\frac{1}{n}\sum_{i=1}^{n} Y_i - E[Y]\right), \qquad (9.17)$$

where c is a constant which can be estimated using the samples (X_i, Y_i) as follows:

$$c = \frac{\sum_{i=1}^{n}(X_i - \bar{X})(Y_i - \bar{Y})}{\sum_{i=1}^{n}(Y_i - \bar{Y})^2}, \qquad (9.18)$$

where $\bar{X}$ and $\bar{Y}$ are the sample mean.

The cautious reader should note that as $n \to \infty$, $\frac{1}{n}\sum_{i=1}^{n} Y_i \to E[Y]$.

Hence, the second term in Eqn. (9.17) evaluates to zero. However, since the number of samples is finite, the samples of Y are going to reduce the variance in the estimator of $E[X]$. The result is an estimator that is better than using only CMC.

As an example, consider the following integral which is to be estimated using control variates:

$$I = \int_0^1 e^x \, dx. \tag{9.19}$$

This integral is the expected value of the function $f(x) = e^x$, where x is a uniform random variable defined over the interval $(0, 1)$.

Let Y be a uniform random variable over the interval $(0, 1)$. The mean of Y is $\frac{1}{2}$ (i.e., $E[Y] = \frac{1}{2}$). Y will be used as the control variate. Listing 9.5 shows the Python implementation of the control variates method for estimating the integral in Eqn. (9.19).

Listing 9.5
Estimating an integral in Eqn. (9.19) using the method of control variates.

```
1   from random import *
2   from math import *
3   from statistics import *
4
5   n = 10000
6
7   Y_mean = 1/2
8
9   X = []
10  Y = []
11
12  for i in range(n):
13      u = random()
14      X.append( exp(u) )
15      Y.append(u)
16
17  X_bar = mean(X)
18  Y_bar = mean(Y)
19
```

```
20   # Auxiliary lists for computing c
21   A = []
22   B = []
23
24   for i in range(n):
25       A.append( (X[i] - X_bar) * (Y[i] - Y_bar) )
26       B.append( (Y[i] - Y_bar)**2 )
27
28   c = sum(A) / sum(B)
29
30   # Samples of CV-based estimator
31   Z = []
32   for i in range(n):
33       Z.append( X[i] - c * ( Y[i] - Y_mean) )
34
35   # Answer using CMC
36   print("I(CMC) = ", round(mean(X), 4), ", Variance = ", round
         (variance(X), 4))
37   # Answer using Control Variates (CV)
38   print("I(CV) = ", round(mean(Z), 4), ", Variance = ", round(
         variance(Z), 4))
39
40   # Output
41   # I(CMC) =  1.7299 , Variance =  0.2445
42   # I(CV) =  1.7185 , Variance =  0.0039
```

9.4.2 Stratified Sampling

The word *stratify* means to *arrange* and *classify*. In this sampling technique, the goal is to stratify samples into groups and then a sample is randomly generated from each group. In this way, samples are spread appropriately across the state space.

Hence, in order to estimate the expected value of a function $f(x)$, the sample space of the random variable X is partitioned into K subsets (i.e., strata) as follows:

$$E[f(x)] = \sum_{i=1}^{K} E[f(x)|x \in S_i] \cdot P[x \in S_i] \tag{9.20}$$

$$E[f(x)|x \in S_i] = \frac{1}{N_i} \cdot \sum_{j=1}^{N_i} f(x_j^i), \tag{9.21}$$

where x_j^i is a sample drawn from the conditional probability distribution $P[x|x \in S_i]$. Hence, the estimator of $E[f(x)]$ using stratified sampling is the following:

$$E[f(x)] = \sum_{i=1}^{K} \frac{1}{N_i} \cdot \sum_{j=1}^{N_i} f(x_j^i) \cdot p_i, \tag{9.22}$$

where $p_i = P[x \in S_i]$, $N_i = p_i \cdot N$, N is the size of the state space S, and $S = \bigcup_{i}^{K} S_i$.

Listing 12.2 shows the implementation of the stratified sampling method in Python. The first part of the program gives the CMC implementation for the purpose of comparing the quality of the two estimators.

Listing 9.6
Estimating the integral $\int_0^1 e^{-x} dx$ using the crude Monte Carlo and stratified methods.

```
1   from random import *

2   from math import *

3   from statistics import *

4

5   n = 10000

6

7   X = []

8

9   for i in range(n):

10      u = random()

11      X.append( exp(-u) )

12

13  print("I(CMC) = ", round(mean(X), 4), ", Variance = ", round
        (variance(X), 4))
```

```
14
15   Y = []
16
17   K = 4     # Number of strata
18   N_i = int(n / K)        # Number of samples from each stratum
19
20   for i in range(K):
21       for j in range(N_i):
22           a = i * 1/K
23           b = a + 1/K
24           u = uniform(a,b)
25           Y.append( exp(-u) )
26
27   print("I(Stratified) = ", round(mean(Y), 4), ", Variance = "
           , round(variance(Y), 4))
28
29   # Output
30   # I(CMC) =   0.6323 , Variance =   0.0325
31   # I(Stratified) =   0.6309 , Variance =   0.0323
```

9.4.3   Antithetic Sampling

This technique was introduced in [4]. The word *antithetic* means opposite. A random variate v has an antithetic value (or variate) that is represented by v'. If v is a random variate uniformly distributed on $[a, b]$, then its antithetic variate is given by

$$v' = a + b - v. \qquad (9.23)$$

The essence of the antithetic sampling technique is to replace each sample s by another one which can be calculated as follows:

$$s^* = \frac{v + v'}{2}. \qquad (9.24)$$

where s' is the antithetic variate of s. Figure 9.7 illustrates how the antithetic value is calculated for each point in the sample space of the random experiment of throwing two dice. Surprisingly, this simple technique leads to a significant reduction in the variance for the same number of samples.

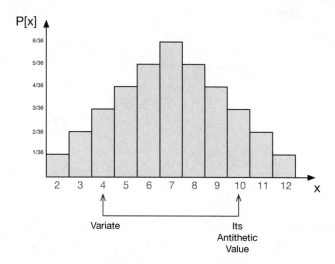

Figure 9.7

Sample space of the random experiment of throwing two dice. For the random variate 4, $\frac{4+10}{2} = 7$ is generated instead if antithetic sampling is used.

Listing 9.7
Estimating the mean of a uniform random variable using antithetic sampling.

```python
from random import *
from statistics import *

n = 1000

# Parameters of the uniform distribution
a = 2
b = 48

# Samples generated using the crude Monte Carlo method
S_cmc = []

# Samples generated using the antithetic method
S_ant = []

for i in range(n):
```

```
17    v = uniform(a, b)
18    S_cmc.append( v )
19    v_ = a + b - v
20    S_ant.append( (v + v_) / 2 )
21
22  print('Mean(S_cmc) = ', round(mean(S_cmc), 4), ", Variance =
          ", round(variance(S_cmc), 4))
23  print('Mean(S_ant) = ', round(mean(S_ant), 4), ", Variance =
          ", round(variance(S_ant), 4))
24
25  # Output
26  # Mean(S_cmc) =   25.6361 , Variance =   178.0452
27  # Mean(S_ant) =   25.0 , Variance =   0.0
```

Listing 9.8 shows how the value of the following integral can be estimated using antithetic sampling.

$$\int_0^1 e^{x^2}\, dx. \tag{9.25}$$

Although both the crude Monte Carlo and antithetic sampling methods achieve a good accuracy, they significantly differ in the variance. Antithetic sampling achieves a very low variance for the same number of samples.

Listing 9.8
Estimating the value of the integral in Eqn. (9.25) using CMC and antithetic sampling. The reduction in variance is about 12%.

```
1  from random import *
2  from statistics import *
3  from math import *
4
5  n = 10000
6
7  S_cmc = []
8  S_ant = []
9
```

```
10  for i in range(n):

11      u = random()

12      u_ = 1 - u

13      S_cmc.append( exp(u**2) )

14      S_ant.append( ( exp(u**2) + exp(u_**2) ) / 2)

15

16  print("Mean(S_cmc) = ", round(mean(S_cmc), 4), ", Variance =
           ", round(variance(S_cmc), 4))

17  print("Mean(S_ant) = ", round(mean(S_ant), 4), ", Variance =
           ", round(variance(S_ant), 4))

18

19  # Output

20  # Mean(S_cmc) =   1.4693 , Variance =   0.2296

21  # Mean(S_ant) =   1.4639 , Variance =   0.0287
```

9.4.4 Dagger Sampling

In dagger sampling, multiple samples can be generated using a single random number. As shown in Figure 9.8, three samples of the random variable X are generated using a single random number u. X has to be a Bernoulli random variable. This is required in order to use this sampling procedure. The number of trials (i.e., samples) is equal to $S = \lfloor \frac{1}{p} \rfloor$, where p is the success probability of the Bernoulli random variable.

Dagger sampling works as follows. The interval $[0, 1]$ is divided into S subintervals. The length of each subinterval is equal to p. The remaining part beyond all the subintervals is not considered. In the example shown in Figure 9.8, there are three subintervals. The random value $u = 0.4$ falls in the second subinterval, which corresponds to the second trial. Hence, the second sample is 'H' and the other samples are all 'T'.

Listing 9.9 shows how dagger sampling is used for estimating the reliability of the system in Figure 9.6(b). In each trial, three samples are initialized on lines 21-23. These three samples are generated as shown on lines 24-31. At the end of each trial, the three samples are checked if they correspond to a connected system.

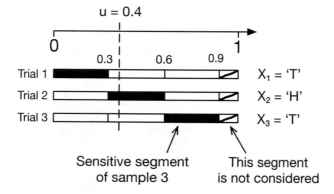

Figure 9.8
With dagger sampling, three trials are performed using a single random number. Hence, three samples are generated.

```
Listing 9.9
Estimating the reliability of the system in Figure 9.6(b) using dagger sampling.

1   from random import *

2   from math import *

3

4   Num_Trials = 10000

5   count = 0

6   p = 0.3     # Probability a block is working

7

8   S = floor(1 / p) # No. of subintervals (samples)

9

10  def Phi(X):

11    if sum(X) == 3:

12      return 1

13    else:

14      return 0

15
```

```
16   # Three samples are generated in each iteration
17   # Total number of samples is S * Num_Trial
18   Total_Num_Samples = S * Num_Trials
19
20   for i in range(Num_Trials):
21       s1 = [0] * 3
22       s2 = [0] * 3
23       s3 = [0] * 3
24       for j in range(3):
25           u = random()
26           if u <= p:
27               s1[j] = 1
28           elif p < u <= 2*p:
29               s2[j] = 1
30           elif 2*p < u <= 3*p:
31               s3[j] = 1
32
33       count = count + Phi(s1)
34       count = count + Phi(s2)
35       count = count + Phi(s3)
36
37   print('Rel_sys = ', round(count / Total_Num_Samples, 3))
38
39   # Output
40   # Exact = 0.027
41   # Rel_sys =  0.028
```

9.4.5   Importance Sampling

In importance sampling, samples are generated using a new probability distribution q that is more appropriate than the original probability distribution p. However, since the new probability distribution q is different from the correct probability distribution p, a correction step is necessary.

Consider the function $g(x)$ in Figure 9.9(a) which is a function of the

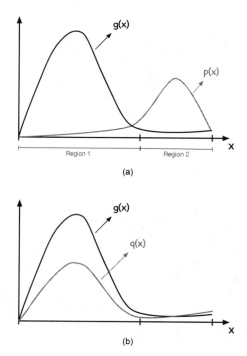

Figure 9.9
Values of $g(x)$ are very close to zero over region 1. Probability distribution of x is very close to zero over region 1. Another probability distribution has to be used in order to generate samples from region 1 where the values of the function $g(x)$ are more interesting.

random variable x whose probability distribution is given by $p(x)$. Values of the function in region 2 will be generated more frequently because of the large probabilities over this region. However, the values of the function in region 1 are more important. How can we sample more frequently from this region? This is the reason why this method is referred to as importance sampling.

Clearly, values of the function $g(x)$ in region 1 have a greater impact on the output (i.e., computed average). Hence, these values should be sampled more frequently. It is, however, very hard to generate samples from region 1 because the $p(x)$ is very small over this region. Thus, we have to use another probability distribution like the one in Figure 9.9(b). This new probability distribution emphasizes the region where the values of the function are more interesting. As a result, a correction step is needed.

It turns out that this correction step is very simple. Basically, every sample generated using $q(x)$ is multiplied by a weight $w(x) = \frac{p(x)}{q(x)}$ to account for the bias that was intentionally introduced. $w(x)$ is referred to as the importance weight.

Next, we mathematically show how the new probability distribution $q(x)$ is introduced into the equation used for computing the average of $g(x)$. The final expression is different because it includes a new term which is $w(x)$.

$$
\begin{aligned}
E[g(x)] &= \sum_{i=1}^{m} g(x_i) \cdot p(x_i) \\
&= \sum_{i=1}^{m} g(x_i) \cdot \frac{p(x_i)}{q(x_i)} \cdot q(x_i) \\
&= \sum_{i=1}^{m} g(x_i) \cdot w(x_i) \cdot q(x_i) \\
&= \sum_{i=1}^{m} g(x_i) \cdot q(x_i) \cdot w(x_i).
\end{aligned}
$$

Computationally speaking, the new notation for the average of $g(x)$ computed using importance sampling is the following:

$$
E[g(x)] \approx \frac{1}{N} \sum_{i=1}^{N} g(x_i) \cdot w(x_i), \quad \text{where} \quad x_i \sim q. \tag{9.26}
$$

Listing 9.10 shows how this sampling procedure is used in estimating the expected value of a function of a random variable.

Listing 9.10
Estimating the average of a function using importance sampling.

```
1   from random import *
2
3   N = 100000
4   E_g = 0
5
6   def g(x):
7       return 8*x
8
9   for i in range(N):
10      x = random()   # Sample from p(x)
11      y = normalvariate(0, 10)   # Sample from q(x)
12      w = x/y   # Importance weight for current sample
13      E_g = E_g + g(y) * w
```

```
14
15  print("E[g(x)] = ", round(E_g / N, 2))   # Answer = 4.0
```

9.5   SUMMARY

Monte Carlo is a powerful tool for estimating probabilities and expected values. The reader is reminded that the design of MC algorithms is not as straightforward as one might think. This is specially true in applications containing events with small probabilities (i.e., rare events).

9.6   EXERCISES

9.1   Using the CMC method, write a program for estimating the probability $P[X > 5]$, where X is a Poisson random variable with parameter $\lambda = 2$. Compare the estimated probability with the exact value.

9.2   Consider the network in Figure 9.10 where the length of each edge is a random variable normally distributed over $[1, 5]$. The random variables are IID. Write a program for estimating the expected length of the shortest path between nodes A and D.

9.3   Using the method of control variates, estimate the integral $I = \int_0^2 e^{-x^2} dx$ using an appropriate Y.

9.4   Using importance sampling, write a program for estimating the probability $P[X \in [10, \infty)]$, where X has an exponential distribution with parameter $\lambda = 1$.

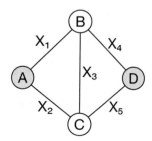

Figure 9.10
Estimating the shortest path between nodes A and D (see Exercise 9.2).

IV

Sources of Randomness

## CHAPTER 10

Random Variate Generation

"It would have been unscientific not to guess."
—Richard Feynman

In a simulation program, an activity typically lasts for an amount of time that is greater than zero. An activity such as the service time is represented by a random variable. Since a random variable is characterized by a probability distribution function, it is this function that is used to generate random variates. A random variate is just a random number generated according to a specific probability distribution. It can also be referred to as a sample or observation. In this chapter, you are going to learn about generating random variates from probability distributions. Some specialized methods are also covered.

10.1 THE INVERSION METHOD

Before delving into the details of the inversion method, let us first describe it at a high level. Remember that a random variable is a function that takes as an input a numerical value and returns a probability. If this function is inversed, what do you think we would get? We will get a new function that takes as an input a probability and returns a numerical value.

Figure 10.1(a) shows the Cumulative Distribution Function (CDF) of an exponential random variable. For every number on the x-axis, there must be exactly one number on the y-axis. Similarly, for every number on the y-axis, there must be exactly one number on the x-axis. Clearly, there is a one-to-one (u-to-v) relationship which can be reversed. The result is the inverse CDF (iCDF) for generating random variates from the exponential distribution. The same reasoning applies to all continuous random variables.

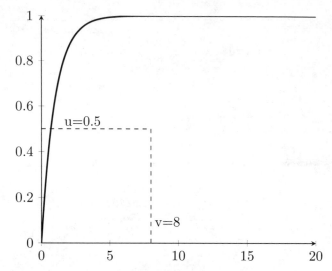

(a) A continuous CDF where $u = 0.5$ corresponds to $v = 8$ ($u = 1 - e^{-v}$).

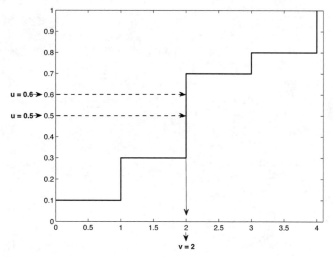

(b) Multiple random numbers are mapped onto one random variate in a discrete CDF.

Figure 10.1
Generating random variates from cumulative distribution functions.

On the other hand, the inversion method works differently on discrete random variables. Figure 10.1(b) shows the CDF of a discrete random variable. In this case, the relationship is many-to-one. That is, multiple random numbers can be mapped onto one random variate. This only applies to the iCDF of discrete random variables.

Finally, it should be emphasized that only CDFs of continuous random variables and Probability Mass Functions (PMFs) of discrete random variables can be inversed. This is because these two functions are actual probability functions. The Probability Density Function (PDF) of a continuous random variable is not a probability function since it can take values greater than one. However, PDFs can be used in the rejection method.

10.1.1 Continuous Random Variables

The process of finding the Random Variate Generator (RVG) of a continuous random variable is systematic. Next, three examples will be discussed in detail to demonstrate this process.

Example 10.1: The Uniform RVG

1. Start with the CDF of the uniform random variable.

$$F\left(x\right) = \frac{x-a}{b-a}.$$

2. Replace $F(x)$ with y.

$$y = \frac{x-a}{b-a}.$$

3. Swap x and y.

$$x = \frac{y-a}{b-a}.$$

4. Solve for y.

$$x(b-a) = (y-a)$$
$$xb - xa = y - a$$

$$xb - xa + a = y$$

$$y = x(b-a) + a.$$

5. Replace y with v and x with u.

$$v = u \cdot (b-a) + a.$$

6. The following expression is used as the uniform RVG:

$$\boxed{v = u \cdot (b-a) + a}$$

Example 10.2: The Exponential RVG

1. Start with the CDF of the exponential random variable.

$$F(x) = 1 - e^{-\mu x}.$$

2. Replace $F(x)$ with y.

$$y = 1 - e^{-\mu x}.$$

3. Swap x and y.

$$x = 1 - e^{-\mu y}.$$

4. Solve for y.

$$e^{-\mu y} = 1 - x.$$

$$ln(e^{-\mu y}) = ln(1 - x).$$

$$-\mu y = ln(1 - x).$$

$$y = \frac{-1}{\mu} \cdot ln(1 - x).$$

5. Replace y with v and x with u.

$$v = \frac{-1}{\mu} \cdot ln(1 - u).$$

$1 - u$ can be replaced with u only.

6. The following expression is used as the exponential RVG:

$$\boxed{v = \frac{-1}{\mu} \cdot ln(u)}$$

Example 10.3

1. Start with the CDF of the random variable.

$$F(x) = \frac{x}{x + 1}; x > 0.$$

2. Replace $F(x)$ with y.

$$y = \frac{x}{x + 1}.$$

3. Swap x and y.

$$x = \frac{y}{y+1}.$$

4. Solve for y.

$$x(y+1) = y$$

$$xy + x = y.$$

$$x = y - xy$$

$$x = y(1-x).$$

$$y = \frac{x}{1-x}.$$

5. Replace y with v and x with u to get the RVG:

$$\boxed{v = \frac{u}{1-u}}$$

10.1.2   Discrete Random Variables

Consider a PMF that models a random experiment having n outcomes. The following is the definition of this PMF.

$$P(X = i) = p_i, i = 0, 1, 2, ..., (n-1), n \in \mathbb{N}^+. \tag{10.1}$$

The CDF for the above PMF can be expressed as the following:

$$F(i) = P(X \le i) = \sum_{j=0}^{i} p_j. \tag{10.2}$$

Hence, the RVG for a discrete random variable can be described as follows:

$$v = \begin{cases} 0 & \text{if } 0 \le u < p_0, \\ 1 & \text{if } p_0 \le u < p_0 + p_1 \ (= \sum_{j=0}^{1} p_j), \\ 2 & \text{if } \sum_{j=0}^{1} p_j \le u < \sum_{j=0}^{2} p_j, \\ ... & \\ ... & \\ (n-1) & \text{if } \sum_{j=0}^{n-2} p_j \le u < \sum_{j=0}^{n-1} p_j, \end{cases} \tag{10.3}$$

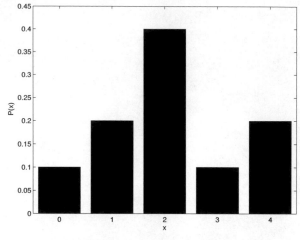

(a) Probability mass function.

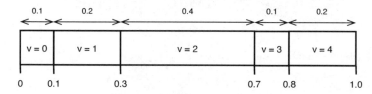

(b) Distribution of mass along a string of one unit of length.

Figure 10.2
Generating random variates using the PMF of a discrete random variable.

where v is the random variate and u is a random number in $(0,1)$.

It should be pointed out that when translating Eqn. (10.3) into code, every condition should be translated to one comparison statement as shown on lines 7, 9, 11, and 13 in Listing 10.1. Notice that the probabilities used on line 3 are the individual probabilities for each possible value of the random variable. However, the probabilities used in the conditions of the if-statement are the cumulative probabilities, which are computed as shown in Figure 10.2(b). In this figure, the width of each interval is still equal to the raw probability of the random variate it represents. However, the five intervals are placed along the x-axis between zero and one to cover the range of possible values of the random number u.

Listing 10.1
Generating random variates using the information in Figure 10.2(a).

```
1  import random as rnd
```

```
 2
 3   p = [0.1, 0.2, 0.4, 0.1, 0.2]
 4
 5   u = rnd.random()
 6
 7   if (0 <= u < p[0]):
 8       v = 0
 9   elif (p[0] <= u < sum(p[0:2])):
10       v = 1
11   elif (sum(p[0:2]) <= u < sum(p[0:3])):
12       v = 2
13   elif (sum(p[0:3]) <= u < sum(p[0:4])):
14       v = 3
15   else:
16       v = 4
17
18   print ('u = ' , u , ' , v = ' , v)
```

10.1.2.1   Generating a Bernoulli Variate

A Bernoulli random variable is a discrete random variable that represents an experiment having two outcomes only. The outcome is either a success with probability p or a failure with probability $1 - p$. Listing 10.2 shows how to generate random variates from Bernoulli probability distribution.

Listing 10.2
Generating Bernoulli random variates.

```
 1   import random as rnd
 2
 3   p = 0.5            # Probability of success
 4
 5   u = rnd.random()
 6
```

```
7   if 0 <= u <= p:
8       print ( '1' )      # Success
9   else:
10      print ( '0' )      # Failure
```

10.1.2.2 Generating a Binomial Variate

A binomial random variable is a discrete random variable that represents the number of successes in a sequence of n independent Bernoulli experiments. When $n = 1$, the binomial random variable becomes a Bernoulli random variable. Listing 10.3 shows how a binomial random variate can be generated using Python.

Listing 10.3
Generating binomial random variates.

```
1   import random as rnd
2
3   p = 0.3                 # Probability of success
4   n = 10                  # Number of trials
5   count = 0               # Count number of successes
6
7   def Bernoulli(p):       # Bernoulli RVG Function
8       u = rnd.random()
9       if 0 <= u < p:
10          return 1
11      else:
12          return 0
13
14  for i in range(n):
15      count = count + Bernoulli(p)
16
17  print ( 'v = ' , count )
```

10.1.2.3 Generating a Geometric Variate

A geometric random variable is a discrete random variable that represents the number of Bernoulli trails needed before getting a success. That is, it represents the number of failures before the first success. For example, you can use this random variable to model a situation where you want to know how many transmission attempts are to be performed before a packet is successfully delivered to the destination. Listing 10.4 shows how to simulate the geometric random variable in Python.

Listing 10.4
Generating geometric random variates.

```python
import random as rnd

p = 0.6    # Probability of success
# Number of Bernoulli trials Needed Before the First Success
count = 0

def Bernoulli(p):
    u = rnd.random()
    if 0 <= u < p:
        return 1
    else:
        return 0

while(Bernoulli(p) == 0):
    count = count + 1

print( 'v = ' , count )
```

10.2   THE REJECTION METHOD

The inversion method fails if you do not have a closed-form expression for the CDF. You can still approximate the CDF using some numerical techniques. However, such techniques often require a significant amount of computational time. Because of these two reasons, the rejection method was invented.

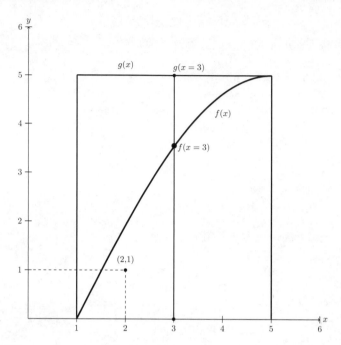

Figure 10.3
Generating a random variate from the PDF f(x) using the auxiliary PDF g(x). $x = 3$ is accepted if $u \leq \frac{f(3)}{g(3)}$.

In this method, the PDF of the random variable (i.e., $f(x)$) is used instead of its CDF. In addition, another auxiliary PDF $g(x)$ is used. Only one assumption is made about $g(x)$. That is, we know how to generate a random variate using $g(x)$. Thus, $g(x)$ can be as simple as the uniform PDF which has a rectangular shape.

Figure 10.3 shows how $g(x)$ can be used to enclose $f(x)$. In fact, any PDF with a very complicated shape can always be enclosed within a uniform PDF. In this way, two regions are created. The first one is enclosed by the curve of $f(x)$. The second one is above the curve of $f(x)$ and below that of $g(x)$. Now, when a point (x, y) is randomly generated such that $1 \leq x \leq 5$ and $0 \leq y \leq 5$, we can visually tell if it lies below the curve of $f(x)$ or above it. If it lies below the curve, then the x-coordinate of the point is reported as the random variate generated according to $f(x)$, which is what we are after. But, how do we generate the random point (x, y)?

The coordinates are generated as uniform random variates. x is uniformly distributed between 1 and 5. Also, y is uniformly distributed between 0 and 5. x is accepted if $y \leq f(x)$, where $y = u \cdot g(x)$. It is this condition that will make sure that the generated random variates will follow the distribution of $f(x)$.

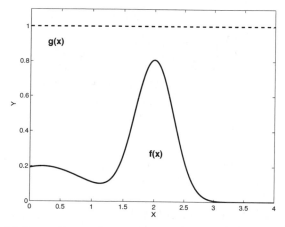

(a) Computing random variates using $g(x) \sim U(0,1)$ and $X \sim U(0,4)$.

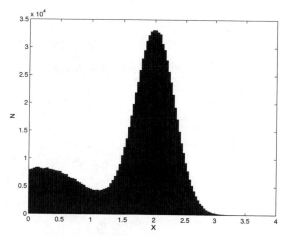

(b) The histogram of one million random variates generated using the rejection method. The shape of the histogram matches that of $f(x)$.

Figure 10.4
Random variates generated using the rejection method.

The following example further illustrates the rejection method. Consider the following PDF:

$$f(x) = 0.2 \times e^{-(x-0.2)^2} + 0.8 \times e^{\frac{-(x-2)^2}{0.2}}.$$

We want to generate random variates such that they follow $f(x)$. Figure 10.4(a) shows the shape of $f(x)$. It also shows $g(x)$ which encloses $f(x)$. In this example, since the maximum value of $f(x)$ is 0.8, we choose $g(x) = 1$ for all $x \in [0, 4]$. $g(x)$ is always a constant and it can be greater than one.

Listing 10.5 shows a procedure for computing random variates from $f(x)$. The functions $f(x)$ and $g(x)$ are declared on lines four and eight, respectively. Then, on line 12, the random variate generation process starts. First, x is randomly assigned a value from its set of values. Then, a uniform number is generated and then used in the comparison on line 15. Notice that $u \times g(x)$ represents a value on the y-axis. If this value is less than or equal to the value of $f(x)$ at the same x, then the point (x, y) lies below the curve of $f(x)$ and x can be accepted as a random variate. Otherwise, the process is repeated. In fact, $\frac{f(x)}{g(x)}$ is a probability (i.e., $0 \leq \frac{f(x)}{g(x)} \leq 1$).

In order to check the validity of the procedure in Listing 10.5, one million random variates are generated and then a histogram is constructed. Figure 10.4(b) shows the histogram. Clearly, the distribution of the generated random variates follows that of $f(x)$.

Listing 10.5
Generating random variates based on the rejection method.

```python
import random as rnd
import math as M

def f(x):
    return 0.2 * M.exp(-(x - 0.2)**2.0) +
                0.8 * M.exp(-(x - 2.0)**2.0 / 0.2)

def g(x):
    return 1      # Uniform PDF

Stop = False
while not Stop:
    x = rnd.uniform(0, 4)        # Generate x
    u = rnd.random()             # y = u * g(x)
    if u <= f(x) / g(x):         # y <= f(x)
        print x
        Stop = True
```

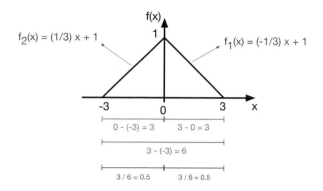

Figure 10.5
Triangular distribution. $f(x)$ is composed of two functions each of which is defined over half the interval over which $f(x)$ exists.

10.3 THE COMPOSITION METHOD

An interesting fact is that a linear combination of CDFs, PDFs, or PMFs is also a CDF, PDF, or PMF, respectively. The only requirement is that the weights used in the combinations should add up to one. Hence, a probability distribution can be represented as a mixture (i.e., weighted sum) of simpler probability distribution functions.

The composition method works as follows. First, the probability function is decomposed into a weighted sum of K simpler probability functions.

$$f(x) = \sum_{i=1}^{K} p_i f_i(x)$$

$$= p_1 f_1(x) + p_2 f_2(x) + \ldots + p_K f_K(x).$$
(10.4)

The condition $\sum_{i=1}^{K} p_i f_i(x) = 1$ must be satisfied.

Secondly, one of the probability distributions that appear in the composition is randomly selected selected. f_i is selected with probability p_i. Finally, a sample is generated using the selected probability distribution function by using either the inversion or rejection method. Clearly, two random numbers are needed. The first one is for choosing the probability distribution function and the second one is for generating a random variate from the selected function.

Example 10.4 illustrates the composition method by showing how a sample can be generated from a triangular distribution shown in Figure 10.5. Clearly, the probability function $f(x)$ can be decomposed into two functions as follows:[1]

---

[1]A line has an equation of the form $y = m \cdot x + b$, where $m = \frac{\triangle y}{\triangle x}$ (slope) and b is the y-intercept.

$$f(x) = \begin{cases} \frac{-1}{3} \cdot x + 1 & \text{if } 0 \le x \le 3, \\ \frac{1}{3} \cdot x + 1 & \text{if } -3 \le x < 0, \\ 0 & \text{otherwise.} \end{cases} \tag{10.5}$$

Example 10.4: Sampling from a triangular distribution.

1. Write $f(x)$ as a composition of two functions:

$$f(x) = p1 \cdot f_1(x) + p2 \cdot f_2(x)$$

where

$$f_1(x) = \frac{-1}{3} \cdot x + 1,$$

and

$$f_2(x) = \frac{1}{3} \cdot x + 1.$$

2. Determine p_1 and p_2. From Figure 10.5, the size of the interval over which each function is defined is 3. The size of the interval over which the original function f is defined is 6. Hence, each function is defined over an interval that represents 50% of the domain of the original function.

$$p_1 = p_2 = 0.5.$$

3. Integrate each function to obtain its CDF. The constant of integration can be calculated using one of the following two points: $(-3, 0)$ or $(3, 1)$. In this example, the latter is used.

$$F_1(x) = \frac{-1}{6} \cdot x^2 + x - \frac{1}{2}.$$

$$F_2(x) = \frac{1}{6} \cdot x^2 + x - \frac{7}{2}.$$

4. Apply the inversion method.

$$F_1^{-1}(u) = 3 - \sqrt{6 - 6u}$$

$$F_2^{-1}(u) = \sqrt{6u + 30} - 3$$

5. Now, given a random number u, a sample v is generated as follows:

$$v = \begin{cases} 3 - \sqrt{6 - 6u} & \text{if } u \le 0.5, \\ \sqrt{6u + 30} - 3 & \text{if } u > 0.5. \end{cases}$$

10.4   THE CONVOLUTION METHOD

Consider a random variable Y whose probability distribution is complex and thus we cannot sample from it. However, this random variable can be expressed as a sum of K random variables $(X_1, X_2, ..., X_K)$ whose probability distributions can be different but they are easy to sample from. In this case, the convolution method can be used to generate samples from Y as follows:

$$Y = X_1 + X_2 + ... + X_K. \tag{10.6}$$

Hence, a sample of Y is the sum of the samples $x_1, x_2, ..., x_K$.

$$y = x_1 + x_2 + ... + x_K.$$

Listing 10.6 shows how an Erlang variate can be generated using the convolution method. Remember that an Erlang random variable is a sum of k independent exponential random variables. For more details, see Section 4.2.7. Figures 10.6(a) and 10.6(b), respectively, show the PDF of the Erlang random variable and histogram constructed from the Elrang random variates generated using the convolution method. Clearly, the two graphs match.

Listing 10.6
Generating an Erlang random variate using the convolution method.

```python
from random import *
from math import *

k = 10
theta = 1.5

y = 0

for i in range(k):
    u = random()
    x = (-1 / theta) * log(1-u)      # Exponential variate
    y = y + x

print("Y", y)
```

A sample from a standard normal random variable with a mean of zero

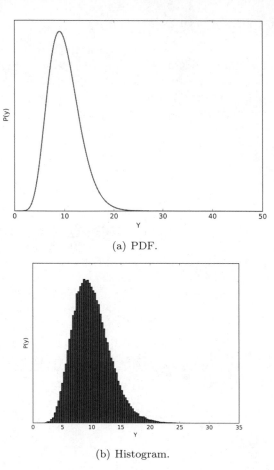

(a) PDF.

(b) Histogram.

Figure 10.6
The shape of the histogram constructed using the random variates generated using the convolution method resembles that of the PDF of the Erlang random variable.

and variance of 1 can also be generated using the convolution method. The reader should be reminded of the following properties of the uniform random variable defined on $(0, 1)$:

1. From Eqn. (4.22) and for $a = 0$ and $b = 1$,

$$\mu = \frac{1}{2},$$

2. From Eqn. (4.23),

$$\sigma^2 = \frac{1}{12}.$$

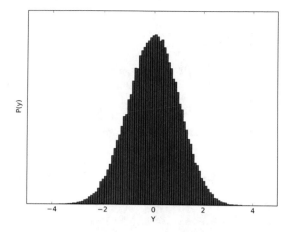

Figure 10.7
Histogram constructed from standard normal variates generated using the convolution method in Listing 10.7.

Therefore, in order to have a variance of one, 12 uniform random variables should be used. Further, in order to have a mean of zero, six must be subtracted from the sum of the 12 uniform random variables. The detail of this procedure is given in Listing 10.7. Figure 10.7 shows the histogram constructed from the standard normal variates generated using the convolution method. Clearly, the bell-shaped graph centered at zero confirms that the random variates are generated from a standard normal probability distribution.

Listing 10.7
Generating a standard normal random variate using the convolution method.

```
1  from random import *
2
3  z = -6
4  for i in range(12):
5      u = random()
6      z = z + u
7
8  print("z = ", z)
```

10.5 SPECIALIZED METHODS

10.5.1 The Poisson Distribution

The Poisson distribution is typically used to model the arrivals in a communication system. Figure 10.8 shows five arrivals during the first time slot in a time-slotted system. Assume that the length of the time slot is one time unit. We know that the inter-arrival time follows the exponential distribution. Thus, every T_i is an exponential random variable with a mean equal to λ. Since the size of the slot is one time unit, the following condition must be met.

$$T_1 + T_2 + T_3 + T_4 + T_5 \leq 1. \tag{10.7}$$

Basically, this condition states that the sum of the five random inter-arrival times must not exceed the length of the time slot. Now, from Example 10.2, the following equation can be used as the random variate generator for T_i.

$$T_i = \frac{-1}{\lambda} \cdot ln(u_i).$$

Hence, Eqn. (10.7) can be re-written as follows.

$$\sum_{i=1}^{5} \frac{-1}{\lambda} \cdot ln(u_i) \leq 1. \tag{10.8}$$

After some algebraic manipulation,[2] the following expression can be obtained.

$$\prod_{i=1}^{5} u_i \geq e^{-\lambda}. \tag{10.9}$$

Inequality (10.9) states that five arrivals can be reported to have occurred if we can sequentially generate five random numbers whose product is greater than or equal to $e^{-\lambda}$. As shown in Figure 10.8, the sixth arrival cannot be considered to have arrived during the first time slot because the sum of the six inter-arrival times would be greater than 1. This is the stopping condition that should be used in the random variate generation scheme for the Poisson distribution.

Listing 10.8 shows how a Poisson random variate can be generated in Python. The left-hand side of inequality (10.9) is implemented by line number 12. The right-hand side, however, is the stopping condition of the while loop (line number 10). So, as long as the product of the generated random numbers does not exceed the threshold set on line number 7, the variable *count* is incremented in every iteration of the while loop. Once the product of the generated random numbers exceeds the threshold, the while loop is exited and the content of the variable *count* is reported as the Poisson random variate.

---

[2]Use the following rules $ln(a \cdot b) = ln(a) + ln(b)$ and $e^{ln(x)} = x$.

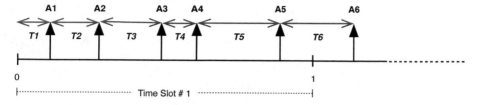

Figure 10.8
Arrivals during a time slot can be modeled as a Poisson random variable.

```
1   import random as rnd
2   import math
3
4   lmda = 10 # Arrival Rate
5   count = 0 # Number of Arrivals
6
7   b = math.exp(-lmda)
8   u = rnd.random()
9
10  while u >= b:
11     count = count + 1
12     u = u * rnd.random()
13
14  print ('v = ' , count)
```

Listing 10.8
Generating a Poisson random variate.

Figure 10.9(a) shows the graph of the Poisson distribution using its PMF with $\lambda = 10$. On the other hand, Figure 10.9(b) shows a histogram of one million Poisson random variates generated using the program in Listing 10.8. In this figure, the $y - axis$ represents the number of variates in every bin of the histogram. To obtain the probability of the Poisson variate corresponding to every bin, the size of the bin is divided by the total number of generated variates. But, clearly, the shape of the histogram resembles that of the PMF of the Poisson random variable with $\lambda = 10$.

10.5.2 The Normal Distribution

Listing 10.9 shows the Python implementation of the Box-Muller method for generating normally distributed random variates from uniformly distributed random numbers as shown in Figure 10.10. The details of this method can be found in [1]. Basically, given two independent random numbers u_1 and u_2, the following equations can be used to generate two independent random variates with a standard normal distribution.

$$z_1 = \sqrt{-2 \cdot ln(u_1)} \cdot cos(2\pi \cdot u_2)$$

$$z_2 = \sqrt{-2 \cdot ln(u_1)} \cdot sin(2\pi \cdot u_2).$$

Figure 10.10 shows in the second row the histograms of u_1 and u_2 before

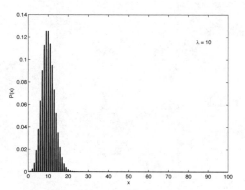

(a) The PMF of the Poisson random variable with $\lambda = 10$.

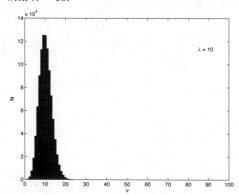

(b) The histogram of one million Poisson random variates generated using the procedure in Listing 10.8 with $\lambda = 10$.

Figure 10.9

The shape of the histogram constructed using simulated Poisson random variates resembles that of the PMF of a Poisson random variable.

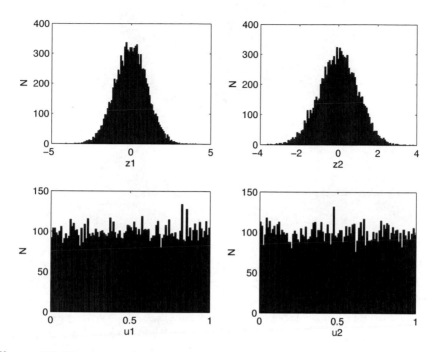

Figure 10.10
Using the procedure in Listing 10.9, the uniform random numbers are transformed into random standard normal variates.

the transformation. The first row, however, shows the resulting histograms of z_1 and z_2 after applying the Box-Muller transformation. Clearly, the shape of the new histograms follows that of the standard normal distribution.

In order to generate a random variate from a non-standard normal distribution with $\mu \neq 0$ and $\sigma \neq 1$, the following equation can be used. In this equation, z is a random variate from a standard normal distribution:

$$v = \mu + \sigma \times z.$$

Listing 10.9
Generating a random variate from a standard normal distribution.

```
from math import sqrt, log, sin, cos, pi
from random import random

def normal(u1,u2):
    z1 = sqrt( -2 * log(u1) ) * cos ( 2 * pi * u2 )
```

```
6      z2 = sqrt ( -2 * log(u1) ) * sin (2 * pi * u2 )
7      return z1 , z2
8
9   u1 = random()
10  u2 = random()
11  z = normal(u1,u2)
12
13  print ('z1 = ' , z[0] , 'z2 = ', z[1])
```

10.6   SUMMARY

Techniques for generating random variates from many important continuous and discrete probability distributions have been introduced and illustrated by examples. The correctness of these techniques have been proved by comparing the shapes of the histograms constructed from the generated random variates with the shapes of the theoretical probability distribution functions.

10.7   EXERCISES

10.1   Consider the following triangular density function defined on $[-1, 1]$:

$$b_i = \begin{cases} 1 + x, & \text{if } -1 \leq x \leq 0 \\ 1 - x, & \text{if } 0 < x \leq 1 \\ 0, & \text{otherwise.} \end{cases}$$

a.   Draw $f(x)$.

b.   Develop a Python program to generate samples from this distribution using the inversion, rejection, composition, and convolution methods.

c.   For each method, generate 10000 random variates and plot the histogram. Does the shape of the histogram match that of the given PDF?

10.2   Write a Python program for generating random variates from the log-normal probability distribution. Use the fact that the natural logarithm of a log-normal random variable has a normal distribution.

## CHAPTER 11

Random Number Generation

"You can recognize truth by its beauty and simplicity."
−Richard Feynman

Random numbers are used in the generation of random variates. A random number u is uniformly distributed between 0 and 1 (denoted by $u \sim U(0,1)$). In this chapter, we are going to describe some popular methods used in the generation of random numbers. Also, we are going to briefly discuss how the performance of these methods can be assessed. Several statistical tests for determining if the generated random numbers really follow the theoretical uniform distribution are covered.

11.1   PSEUDO-RANDOM NUMBERS

The word *pseudo* means not authentic (i.e., false). It is used with the word *random* to mean that a number has a close resemblance to a true random number. This resemblance is confirmed using standard statistical tests.

The program (or device) used to generate pseudo-random numbers is referred to as a Random Number Generator (RNG). The behavior of any RNG is deterministic and predictable. The random numbers generated by an RNG must form a uniform distribution when their histogram is constructed.

A true random number u is a random variable with the following probability distribution:

$$f(u) = \begin{cases} 1 & \text{if } 0 < u < 1, \\ 0 & \text{otherwise.} \end{cases} \qquad (11.1)$$

Figure 11.1 shows the graphical representation of the uniform probability distribution. The following are three important statistics of u:

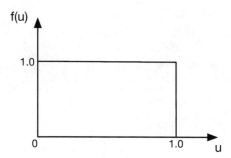

Figure 11.1
Probability distribution of u.

1. Mean

$$E(u) = \frac{1}{2}. \tag{11.2}$$

2. Variance

$$V(u) = \frac{1}{12}. \tag{11.3}$$

3. Expectation of the autocorrelation

$$E(u_i u_{i+1}) = E(u_i)E(u_{i+1})$$

$$= \frac{1}{N} \sum_{i=1}^{N-1} u_i u_{i+1} \tag{11.4}$$

$$= \frac{1}{4}. \tag{11.5}$$

As will be seen later, for a large set of random numbers, the above three statistics can be used as a quick (and first) test for uniform randomness (see Listing 11.1). If the computed values match the above theoretical values, then the generated random numbers could be uniformly distributed. Of course, further testing needs to be done.

Listing 11.1
Testing a set of random numbers if they are uniformly distributed.

```
1  from random import *
2  from statistics import *
3
4  N = 10000
```

```
 5   data = [random() for i in range(N)]

 6

 7   corr = 0

 8   for i in range(N-1):

 9       corr = corr + data[i]*data[i+1]

10   corr = corr / N

11

12   print("Mean = ", round(mean(data), 2))

13   print("Variance = ", round(variance(data), 2))

14   print("Autocorrelation = ", round(corr, 2))

15

16   # Output

17   # Mean = 0.5

18   # Variance = 0.08

19   # Autocorrelation = 0.25
```

11.2   CHARACTERISTICS OF A GOOD GENERATOR

RNGs are the main source of randomness in simulation programs. They are actually programs whose behavior is deterministic. Once its initial state (also called the *seed*) is set, a RNG produces a deterministic and periodic sequence of numbers. This is why we refer to a RNG as a *pseudo*[1] RNG.

A RNG should produce the same sequence of random numbers for the same seed. Only one seed is used in every simulation run. Therefore, the RNG is required to have a long period since the sequence can repeat. Table 11.1 shows two sequences of random numbers generated by two different RNGs. Both RNGs have a cycle of size 3. Because of this, only three values of the random variables IAT and ST are simulated. For example, the effect of having short service times cannot be captured since random variates less than 10 will never occur.

Other desired characteristics of a good RNG are *uniformity* and *independence*. Uniformity means that if the interval $(0, 1)$ is divided into k subintervals of equal length, there will be $\frac{N}{k}$ random numbers in each subinterval, where N is the size of the set of generated random numbers. Independence, on the other hand, means that there is no clear pattern in (or no relationship between) the generated random numbers (e.g., small numbers followed by larger

---

[1]Pseudo means false.

Table 11.1
Random variates are repeated after a cycle of size 3. This is due to the repetition in the generated sequence of the random numbers.

u_1	0.3297	0.6321	0.1813	0.3297	0.6321	0.1813	0.3297	0.6321
IAT	2	5	1	2	5	1	2	5
u_2	0.9093	0.8647	0.9592	0.9093	0.8647	0.9592	0.9093	0.8647
ST	12	10	16	12	10	16	12	10

numbers). To guarantee that all these desirable characteristics are achieved, standard statistical tests are performed (see Section 11.7).

11.3   JUST ENOUGH NUMBER THEORY

11.3.1   Prime Numbers

A prime number is a positive integer that is greater than one and has two divisors only: one and itself. The following numbers are all prime numbers:

$$3, 5, 7, 11, 13, 17, 19, 23, ...$$

Prime numbers are crucial in random number generation. Parameters of RNG algorithms are often recommended to be large prime numbers.

11.3.2   The Modulo Operation

The modulo operation finds the *remainder* of the division of one integer number by another. Given two positive integers a and b, the modulo (also, abbreviated as $a \ mod \ b$) is computed as follows.

$$m = a - \left\lfloor \frac{a}{b} \right\rfloor \cdot b \qquad (11.6)$$

where $b > 0$ and $\lfloor x \rfloor$ denotes the floor function which gives the greatest integer less than or equal to the argument x. This is definition equivalent to the definition of the integer division operation where the fractional part is discarded.

Let us consider an example. If $a = 7$ and $b = 5$, the value of r (i.e., the remainder) can be computed as shown in Example 11.1.

Example 11.1: Calculating $7\ mod\ 5$

$$
\begin{aligned}
r\ &= 7 - \left\lfloor \tfrac{7}{5} \right\rfloor \cdot 5 \\
&= 7 - \lfloor 1.4 \rfloor \cdot 5 \\
&= 7 - (1) \cdot 5 \\
&= 7 - 5 \\
&= 2.
\end{aligned}
$$

11.3.3 Primitive Roots for a Prime Number

For a prime number p, the number b is one of its primitive roots if the set of powers of b; i.e., $\{b^0, b^1, b^2, ...\}$, include all the numbers in the set $\{1, 2, 3, ..., p-1\}$, which is the set of all possible remainders (except zero). Example 11.2 shows that three is a primitive root for seven. All the possible remainders will occur. By contrast, Example 11.3 shows that two is not a root primitive for seven. This is because the numbers $\{3, 5, 6\}$ will never occur as remainders. Two, however, is a root primitive for 13.

Example 11.2: $b = 3$ is a primitive root for $p = 7$.

i	b^i	$b^i \pmod 7$
0	1	1
1	3	3
2	9	2
3	27	6
4	81	4
5	243	5

Example 11.3: $b = 2$ **is not a primitive root for** $p = 7$ **but it is for**
$$p = 13.$$

i	b^i	$b^i \pmod{13}$
0	1	1
1	2	2
2	4	4
3	8	8
4	16	3
5	32	6
6	64	12
7	128	11
8	256	9
9	512	5
10	1024	10
11	2048	7
12	4096	**1**

i	b^i	$b^i \pmod{7}$
0	1	1
1	2	2
2	4	4
3	8	**1**
4	16	2
5	32	4

11.4 THE LINEAR CONGRUENTIAL METHOD

This method is one of the most popular ones. Consider the following relation:

$$X_{n+1} = (a \cdot X_n + c) \mod m, \qquad n \geq 0 \qquad (11.7)$$

where a, c, and m are called the *multiplier*, *increment*, and *modulus*, respectively. The initial number X_0 is referred to as the *seed*. The random number u_n is obtained as follows.

$$u_n = \frac{X_n}{m}, \qquad n \geq 0. \qquad (11.8)$$

Clearly, if a, c, and m are fixed, then different seeds would give different sequences of random numbers. For every simulation run, the seed must be recorded. This is necessary if the simulation run is to be exactly reproduced (i.e., replicated). Changing the seed is referred to as *reseeding* the random number generator (or the simulator). A new seed will give a new sequence of random numbers. This way another path in the system state space is explored (see Figure 11.2).

As an example, consider a linear congruential random number generator with the following paramters: a = 2, c = 3, m = 10, $X_0 = 0$.

n	0	1	2	3	4	5	6	7	8
X_n	0	3	9	1	5	3	9	1	5
u_n	0.0	0.3	0.9	0.1	0.5	0.3	0.9	0.1	0.5

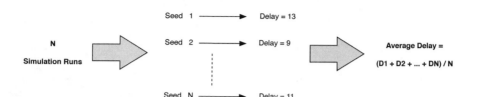

Figure 11.2
Multiple seeds are used to make different simulation runs. Different paths in the system state space are explored and the average is computed using the values resulting from these different paths.

u_0 is not considered as part of the sequence. Thus, the sequence will repeat itself after four steps. That is, the same random number will re-appear after four steps.

11.5   THE MULTIPLICATIVE CONGRUENTIAL METHOD

Another method for generating random numbers is based on the following relation.

$$X_{n+1} = a \cdot X_n \mod m, \qquad n \geq 0 \tag{11.9}$$

where a, m, and X_0 are the multiplier, modulus, and seed, respectively. The random number is then obtained using Eqn. (11.8).

11.5.1 2^k Modulus

In order to produce a long sequence of unique random numbers, the values of the parameters can be set as follows.

X_0 is an odd integer,

a = 8t ± 3, where t is a positive integer, and

m = 2^k, where k is equal to the word size of the computer (e.g., 64 bits).

For a, choose the value which is closest to $2^{\frac{b}{2}}$. If the above recipe is followed, it is guaranteed that we will get a sequence of $2^{(b-2)}$ random numbers before the sequence is repeated.

Consider a multiplicative congruential random number generator with the following paramters: $t = 1$, $b = 4$, and $X_0 = 1$. The resulting sequence will have a period of size four, $a = 11$, and $m = 16$.

n	0	1	2	3	4	5	6	7	8
X_n	1	11	9	3	1	11	9	3	1
u_n	0.062	0.688	0.562	0.188	0.062	0.688	0.562	0.188	0.062

u_0 is not considered as part of the sequence. Thus, the sequence will repeat itself after four steps. That is, the same random number will re-appear after four steps.

11.5.2   Prime Modulus

A multiplicative RNG with a prime modulus will achieve the maximum period; i.e., $M-1$, if the multiplier a is a primitive root for M. Example 11.4 illustrates this relationship.

Example 11.4: $a = 3$, $M = 7$, and $X_{n+1} = 3X_n$ $(mod\ 7)$.

n	0	1	2	3	4	5	6	7
X_n	3	2	6	4	5	1	3	2
u_n	0.43	0.29	0.86	0.57	0.71	0.14	0.43	0.29

u_0 is not considered. Notice that the period of the generated sequence is $M - 1 = 6$, which is the maximum period. Unfortunately, although this generator has a full period, it is a bad one. This is because the period is very short.

A minimal standard generator is proposed in [10]. It has a very long period, which is $2^{31} - 2 = 2,147,483,646$. This generator has the following parameters: $a = 7^5 = 16807$ and $M = 2^{31} - 1$.

11.6   LINEAR FEEDBACK SHIFT REGISTERS

A Linear Feedback Shift Register (LFSR) is a digital device that consists of memory cells and exclusive-OR (XOR) gates. It can generate a sequence of random binary numbers. Figure 11.3 shows a four-bit LFSR that contains only one XOR gate. In LFSRs, XOR gates are inserted between adjacent memory cells using characteristic polynomials of the LFSRs. In order to generate the maximum-length sequence of random numbers, an n-bit LFSR must be constructed using its characteristic polynomial. These are standard polynomials which can be found in standard books on cryptography (e.g., see [9]).

For the four-bit LFSR in Figure 11.3, its characteristic polynomial is $c(x) = 1 + x^3 + x^4$ and it is constructed as follows:

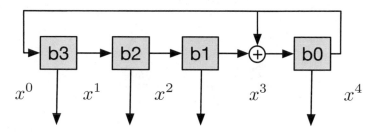

Figure 11.3
A four-bit linear feedback shift register with characteristic polynomial $c(x) = 1 + x^3 + x^4$.

1. Since the characteristic polynomial is of degree $n = 4$, then four memory cells are required,

2. Each term present in the characteristic polynomial (except x^0 and x^n) corresponds to an XOR gate. In this case, an XOR gate is inserted between memory cell one and zero since this place corresponds to x^3, and

3. An initial binary number is loaded into the memory cells (i.e., seed).

Now, the LFSR is ready and random binary numbers can be generated as shown in Table 11.2. The seed is chosen to be 0001. The sequence will repeat when this number occurs again. This LFSR has a period of size 15 ($2^4 - 1$). This is the number of all possible unique random numbers. Clearly, the next random number is predictable if the characteristic polynomial is known.

Listing 11.2 shows how the LFSR in Figure 11.3 can be implemented in Python. Three masks are defined on lines 4-6. The first mask is for extracting the value of the memory cell whose input is the output of the XOR gate (i.e., $b_0 = b_0 \oplus b_1$). A mask is needed for each XOR gate in the LFSR. The second mask is for extracting the values of b_1 and b_2. Finally, the third mask is for extracting b_0.

The next random number is the result of the logical OR operation of three intermediate numbers (see line 16). For instance, the second random number which is 9 is the result of the logical OR operation of $temp1 = 0001$, $temp2 = 0000$, and $temp3 = 1000$ (computed on lines 13-15, respectively). In order to compute the value of $temp1$, two shifted copies of the current random number (i.e., num) are created. The value of b_1 is shifted to the right so that it is aligned with the value of b_0. Then, these two binary strings are combined using an XOR operation. Finally, the new value of b_0 is extracted using its mask (i.e., $mask1$). This process is illustrated in Figure 11.4.

As another example, Listing 11.3 shows the Python implementation of an eight-bit LFSR with characteristic polynomial $c(x) = 1 + x^4 + x^5 + x^6 + x^8$ (see Figure 11.5). Three XOR gates must be inserted before memory cells 3,

Table 11.2
Maximum-length sequence of random numbers generated by the LFSR in Figure 11.3.

b_3	b_2	b_1	b_0	Number
0	0	0	1	1
1	0	0	1	9
0	0	0	1	13
0	0	0	1	15
0	0	0	1	14
0	0	0	1	7
0	0	0	1	10
0	0	0	1	5
0	0	0	1	11
0	0	0	1	12
0	0	0	1	6
0	0	0	1	3
0	0	0	1	8
0	0	0	1	4
0	0	0	1	2
0	**0**	**0**	**1**	**1***

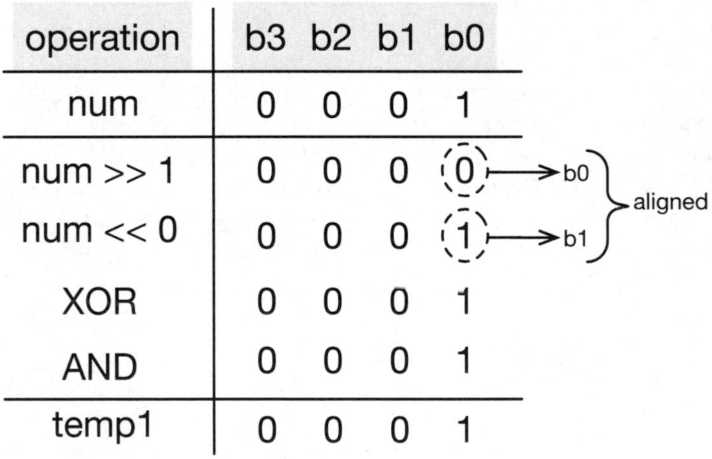

operation	b3	b2	b1	b0
num	0	0	0	1
num >> 1	0	0	0	(0) ⟶ b0
num << 0	0	0	0	(1) ⟶ b1
XOR	0	0	0	1
AND	0	0	0	1
temp1	0	0	0	1

aligned

Figure 11.4
Computing the first intermediate binary number on line 13 in Listing 11.2.

2, and 1. Also, this will be translated to three `bit-shift-left` operations by 1, 2, and 3 bits, respectively (see lines 13-15).

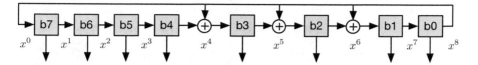

Figure 11.5
An eight-bit linear feedback shift register with characteristic polynomial
$c(x) = 1 + x^4 + x^5 + x^6 + x^8$.

Listing 11.2
Generating the maximum-length random sequence from the four-bit LFSR
shown in Figure 11.3.

```
1   seed = 0b_0001
2   num = seed
3
4   # Define masks
5   mask1 = 0b_0001
6   mask2 = 0b_0110
7   mask3 = 0b_0001
8
9   # Counter
10  period = 0
11
12  while True:
13    print(num)
14
15    temp1 = ( (num >> 1) ^ (num << 0) ) & mask1
16    temp2 = ( num >> 1 ) & mask2
17    temp3 = (num & mask3) << 3
18    num = temp1 | temp2 | temp3
19
20    period += 1
21
22    if num == seed:
23      break
24
```

```
25  print("Period = ", period)

26

27  # Period: 2^4 - 1 = 15
28  # Numbers: 1 9 13 15 14 7 10 5 11 12 6 3 8 4 2
```

Listing 11.3
Generating the maximum-length random sequence from an eight-bit LFSR.

```
1   seed = 0b_00111000
2   num = seed

3

4   # Define masks
5   mask1 = 0b_00000010
6   mask2 = 0b_00000100
7   mask3 = 0b_00001000
8   mask4 = 0b_01110001
9   mask5 = 0b_00000001

10

11  period = 0

12

13  while True:
14      temp1 = ( (num >> 1) ^ (num << 1) ) & mask1
15      temp2 = ( (num >> 1) ^ (num << 2) ) & mask2
16      temp3 = ( (num >> 1) ^ (num << 3) ) & mask3
17      temp4 = ( num >> 1 ) & mask4
18      temp5 = (num & mask5) << 7
19      num = temp1 | temp2 | temp3 | temp4 | temp5

20

21      print(num)

22

23      period += 1

24
```

```
25    if num == seed:
26       break
27
28  print("Period = ", period)
29
30  # Period: 2^8 - 1 = 255
```

11.7   STATISTICAL TESTING OF RNGs

As you can tell by now, the sequence of random numbers generated by any RNG is not truly random! This is because the entire sequence is predictable. Also, since m in equations (11.7) and (11.9) is finite, the sequence will eventually repeat. Fortunately, however, what we require is that the generated sequence has some of the characteristics of a real random sequence. Statistical tests are used to confirm these characteristics.

Typically, a sequence of random numbers is accepted if it satisfies two conditions: *uniformity* and *independence*. Two standard statisitical tests are used for checking these two conditions. The first test is referred to as the *chi-squared test* (χ^2 test). This test ensures that no number occurs more often than the other numbers. This way the numbers are uniformly distributed. The second test which is referred to as the *poker test* ensures that there is no correlation between the successive random numbers. This way the numbers are independent from each other. A RNG is accepted if it passes these two tests. Next, the two tests are described.

11.7.1   The Chi-Squared Test

This test is mainly used for determining how well the observed data (i.e., generated random numbers) fit the theoretically expected data (i.e., uniformly distributed). The test is performed as follows.

1. Divide the interval [0,1) into K non-overlapping subintervals of equal length,

2. Determine O_i for each subinterval i, where O_i is the number of random numbers that fall in subinterval i and $1 \leq i \leq K$,

3. Determine E_i for each subinterval i, where E_i is the expected number of random numbers that should fall in subinterval i, and

4. Compute the chi-squared statistic χ^2 given by the equation

$$\chi^2 = \sum_{i=1}^{K} \frac{(O_i - E_i)^2}{E_i}.$$

5. For a level of significance α, if $\chi^2 \leq \chi^2_{K-1,1-\alpha}$, then it is concluded that the random numbers in the given sequence are uniformly distributed with $((1 - \alpha) \times 100)\%$ level of confidence.

The recommended value for K is at least 10 and $\frac{n}{K}$ should be at least five, where n is the size of the sequence of random numbers. For instance, if $n = 100$ and $K = 10$, we expect that $E_i = 10$, where $1 \leq i \leq 10$. That is what we mean by uniform distribution.

Consider a sequence of random numbers of size 100. If $K = 10$, we expect each subinterval to contain ten random numbers. Assume that the information in column four is obtained after analyzing the given sequence. Then, χ^2 can be calculated as follows:

i	$Range$	E_i	O_i	$(O_i - E_i)^2$	$\frac{(O_i - E_i)^2}{E_i}$
1	$[0, 0.1)$	10	9	1	0.1
2	$[0.1, 0.2)$	10	5	25	2.5
3	$[0.2, 0.3)$	10	12	4	0.4
4	$[0.3, 0.4)$	10	11	1	0.1
5	$[0.4, 0.5)$	10	9	1	0.1
6	$[0.5, 0.6)$	10	8	4	0.4
7	$[0.6, 0.7)$	10	11	1	0.1
8	$[0.7, 0.8)$	10	9	1	0.1
9	$[0.8, 0.9)$	10	10	0	0
10	$[0.9, 1.0)$	10	16	36	3.6
					$\chi^2 = 7.4$

Next, we compute the degree of freedom (df) for the χ^2 statistic. By definition, the degree of freedom is always $K - 1$. After obtaining df, we need to find the row in the chi-squared table corresponding to $df = 9$:

0.995	0.99	0.95	0.90	0.75	0.50	0.25	0.10	0.05	0.01	0.005
1.73	2.09	3.33	4.17	5.90	8.34	11.4	14.7	16.9	21.7	23.6

Assuming a significance level $\alpha = 0.05$, the critical value from the above table is $\chi^2_{9,0.95} = 16.9$. Now, since the obtained value $(\chi^2 = 7.4)$ is less than the critical value, it is concluded that the random numbers in the given sequence are uniformly distributed with a 95% level of confidence.

Table 11.3
Types of five-digit numbers according to the poker rules.

Combination	Type
AAAAA	Five of a Kind
AAAAB	Four of a Kind
AAABB	Full House
AAABC	Three of a Kind
AABBC	Two Pairs
AABCD	One Pair
ABCDE	Five Different Digits

11.7.2 The Poker Test

A sequence of random numbers might be uniformly distributed and yet not random. This is because the random numbers may be related. The poker test is used to detect any such relationship. However, before applying the poker test, the sequence of random numbers must be pre-processed using the following two steps.

1. Remove the decimal point in every random number.

2. Choose the first five digits in every random number. You may need to round the numbers.

Following the above procedure, we will end up with a sequence of five-digit numbers. Now, we are ready to apply the poker test to the random sequence.

   In this test, every random number is treated as a poker hand. Thus, each random number can be classified using the same poker rules. Table 11.3 shows the possible combinations of five-digit numbers that are considered in the poker test. It also shows the type of each combination according to the game of poker.

   Consider a sequence of random numbers of size 100. The following table gives the distribution of the random numbers in the seven possible categories in the poker test.

$Category$	E_i	O_i	$(O_i - E_i)^2$	$\frac{(O_i - E_i)^2}{E_i}$
Five Different Digits	30	35	25	0.83
One Pair	50	51	1	0.02
Two Pairs	10	9	1	0.1
Three of a Kind	7	3	16	2.29
Full House	1	0	1	1
Four of a Kind	1	1	0	0
Five of a Kind	1	1	0	0
				$\chi^2 = 4.24$

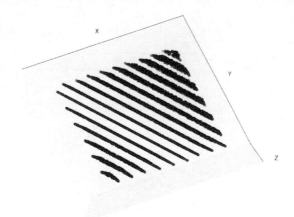

azimuth = -66 deg, elevation = -83 deg

Figure 11.6
10^4 triplets of successive random numbers generated using Listing 11.4. Planes can be seen when the figure is viewed from the right angle.

Notice that the numbers in the second column E_i are based on empirical observations. In fact, they represent percentages and thus can be applied to a random sequence of any length.

The degree of freedom is six since there are seven categories. The critical value of chi^2 for $df = 6$ at $\alpha = 0.05$ is $\chi^2_{6,0.95} = 12.6$. Since the obtained value ($\chi^2 = 4.24$) is less than the critical value, it is concluded that the random numbers in the given sequence are independent with a 95% level of confidence.

11.7.3 The Spectral Test

This test is used for detecting correlations among random numbers. Basically, the random numbers are grouped into triplets. These triplets are plotted in a 3D space. Planes will emerge if the random numbers are correlated. Figure 11.6 shows an example using a random number generator with $a = 65539$ and $M = 2^{31}$. Listing 11.4 is used to produce the figure. You still have to rotate the figure to see the planes. The recommended values for the azimuth and elevation are shown in the figure.

Listing 11.4
Python program for generating a 3D scatter plot for the spectral test.

```
1  import math
```

```
 2   import matplotlib.pyplot as plt
 3   from mpl_toolkits.mplot3d import Axes3D
 4
 5   a = 65539
 6   M = math.pow(2, 31)
 7   seed = 123456
 8
 9   X = []
10   Y = []
11   Z = []
12
13   for i in range(10000):
14       num1 = math.fmod(a * seed, M)
15       num2 = math.fmod(a * num1, M)
16       num3 = math.fmod(a * num2, M)
17       seed = num2
18       X.append(num1)
19       Y.append(num2)
20       Z.append(num3)
21
22   fig = plt.figure()
23   ax = fig.add_subplot(111, projection='3d')
24   ax.scatter(X, Y, Z, c='b', marker='o')
25   # Remove axis ticks for readability
26   ax.set_xticks([])
27   ax.set_yticks([])
28   ax.set_zticks([])
29   ax.set_xlabel('X')
30   ax.set_ylabel('Y')
31   ax.set_zlabel('Z')
32   plt.show()
```

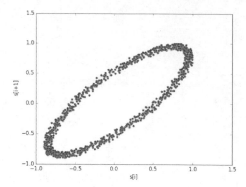

Figure 11.7

The lag plot for a sequence of sinusoidal values. An elliptical pattern can be clearly identified.

11.7.4 The Lag Plot

Given a sequence of random numbers, a lag plot can be used to test if the numbers are really random. Basically, if the lag plot exhibits no patterns, then the numbers are random. On the other hand, if the numbers are not random, the lag plot will show an identifiable pattern, like linear and circular patterns. Figure 11.7 shows an example of the lag plot of a non-random sequence.

The lag plot can be produced using Listing 11.5. The result of running the code is in Figure 11.8. Clearly, in this figure, there is no identifiable pattern. Besides, the random numbers are uniformly distributed in the 2D space. Therefore, the generated sequence is random.

Listing 11.5
Python procedure for generating a lag plot for a random sequence.

```
1  import random as rnd
2  import pandas
3  from pandas.tools.plotting import lag_plot
4  import matplotlib.pyplot as plt
5
6  s = pandas.Series([rnd.random() for i in range(10000)])
7
8  plt.figure()
9  lag_plot(s, marker='o', color='grey')
10 plt.xlabel('Random Number - s[i]')
```

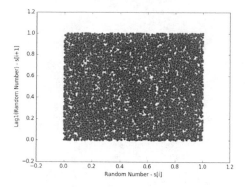

Figure 11.8
The lag plot generated by the code in Listing 11.5. The sequence uniformly fills the 2D space.

```
11  plt.ylabel('Lag1(Random Number) - s[i+1]')
12  plt.show()
```

11.8   SUMMARY

This chapter has discussed several methods for the generation of pseudo-random numbers. These pseudo-random numbers are used in the computation of pseudo-random variates and pseudo-random processes. According to [3], a RNG should not produce a zero or one. In addition, the generated random numbers should look random although they are generated using deterministic procedures.

11.9   EXERCISES

11.1 Show that the multiplicative RNG does indeed pass both the chi-squared and poker tests.

11.2 Consider the 16-bit LFSR with characteristic polynomial $c(x) = 1 + x^4 + x^{13} + x^{15} + x^{16}$. Draw the structure of this LFSR and write a Python program that implements it.

V

Case Studies

## CHAPTER 12

Case Studies

"Discovery is seeing what everybody else has seen and thinking what nobody else has thought."
—Albert Szent-Györgyi

The main purpose of this chapter is to show the reader how the transition from a system description to a simulation model is made. The first case study is about estimating the reliability of a network using the Monte Carlo methods and several variance-reduction techniques. The second case study is about modeling a point-to-point wireless transmission system where packets may be lost either due to a full queue or bad channel state. There is also an upper limit on the number of transmission attempts before the packet is dropped. Both packet delay and system throughput are analyzed. The final case study is about modeling a simple error-control protocol and studying the impact of error probability on system throughput.

12.1 NETWORK RELIABILITY

This case study discusses the reliability evaluation of static networks. A network is referred to as static if time plays no role in its model. A network can be modeled as a graph which consists of vertices (nodes) and edges (links). The nodes are perfect but the links can fail. When links fail, the network becomes disconnected. This is an interesting situation where we can ask the following questions:

1. Is the source still connected to the destination?

2. Is every node reachable from every other node?

3. Are the nodes in a given subset connected?

Figure 12.1 shows a graph $G(V, E)$ of a network that has eight vertices and 11 edges. The set of vertices is $V = \{v_1, v_2, v_3, v_4, v_5, v_6, v_7, v_8\}$. And, the

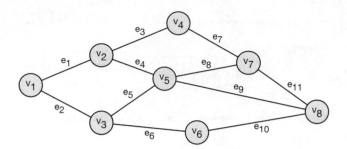

Figure 12.1
A graph consisting of eight vertices and 11 edges.

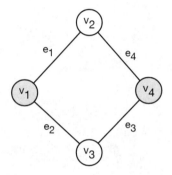

Figure 12.2
Network fails if nodes v_1 and v_4 become disconnected. The event will occur if any of the following groups of links fail: $\{(e_1, e_2), (e_1, e_3), (e_2, e_4), (e_3, e_4)\}$.

set of edges is $E = \{e_1, e_2, e_3, e_4, e_5, e_6, e_7, e_8, e_9, e_{10}, e_{11}\}$. A *path* in a graph is a sequence of edges that start and terminate at two distinct vertices. For example, one path that connects vertex v_1 to v_8 is $\{e_1, e_3, e_7, e_{11}\}$. We say that node v_1 is connected to node v_8 because there is at least one path between them.

Network reliability is defined as the probability that a specific set of nodes in a given graph stay connected while each link can fail independently with probability q. On the other hand, *network unreliability* is the probability of network failure which occurs when the nodes under consideration are not connected. Exact computation of these two metrics is not possible since the runtime grows exponentially with the number of links. Thus, approximate techniques based on Monte Carlo simulation are recommended.

Clearly, from Table 12.1, the unreliability can be calculated using the following expression:

$$UnRel = P[s_1] + P[s_2] + P[s_3] + P[s_4] + P[s_5] + P[s_6] + P[s_9] + P[s_{11}] + P[s_{13}].$$
$$(12.1)$$

Table 12.1
Sample space of the system in Figure 12.2 along with the status of the network for each possible system state.

System State	Status	e_1	e_2	e_3	e_4	$P[s_i]$
s_1	Down	0	0	0	0	q^4
s_2	Down	0	0	0	1	$(1-q)q^3$
s_3	Down	0	0	1	0	$(1-q)q^3$
s_4	Down	0	0	1	1	$(1-q)^2q^2$
s_5	Down	0	1	0	0	$(1-q)q^3$
s_6	Down	0	1	0	1	$(1-q)^2q^2$
s_7	Up	0	1	1	0	$(1-q)^2q^2$
s_8	Up	0	1	1	1	$(1-q)^3q$
s_9	Down	1	0	0	0	$(1-q)q^3$
s_{10}	Up	1	0	0	1	$(1-q)^2q^2$
s_{11}	Down	1	0	1	0	$(1-q)^2q^2$
s_{12}	Up	1	0	1	1	$(1-q)^3q$
s_{13}	Down	1	1	0	0	$(1-q)^2q^2$
s_{14}	Up	1	1	0	1	$(1-q)^3q$
s_{15}	Up	1	1	1	0	$(1-q)^3q$
s_{16}	Up	1	1	1	1	$(1-q)^4$

Similarly, the reliability can be calculated using the following expression:

$$Rel = 1 - UnRel = P[s_7] + P[s_8] + P[s_{10}] + P[s_{12}] + P[s_{14}] + P[s_{15}] + P[s_{16}].$$
$$(12.2)$$

As shown in Listing 12.2, the crude Monte Carlo method generates N realizations of the network and estimates unreliability as the proportion of those realizations in which nodes v_1 and v_4 are disconnected. This can be expressed mathematically as follows:

$$UnRel = \frac{1}{N} \sum_{i=1}^{N} \Phi(s_i),$$

where $\Phi(s_i)$ evaluates to one if the given network realization (i.e., sample) s_i represents a connected network. The alert reader should realize that $\Phi(s_i)$ is a Bernoulli random variable whose expectation is equal to the unreliability of the network.

The crude Monte Carlo method suffers from a fundamental problem. Consider the expression for the expected relative half width confidence interval of u:

$$CI_{hw} = \frac{t \times \frac{s}{\sqrt{n}}}{E[\Phi]}.$$
$$(12.3)$$

Clearly, this expression grows to infinity as u approaches zero for the same

Listing 12.1
Computing unreliability for the graph in Figure 12.2 using the exact expression in Eqn. (12.1).

```
1  q = 0.3
2  Unreliability = q**4 + 4 * (1-q) * q**3 \
3                          + 4 * (1-q)**2 * q**2
4  print("Unreliability = ", round(Unreliability, 10)) # 0.2601
```

Table 12.2
Restructuring the sample space of the system in Figure 12.2 along with the probability of each stratum. The first row indicates the number of *UP* links.

0	1	2	3	4
0000	0001	0011	0111	1111
	0010	0101	1110	
	0100	1001	1101	
	1000	1010	1011	
		1100		
		0110		
$P_0 = 0.0625$	$P_1 = 0.25$	$P_2 = 0.375$	$P_3 = 0.25$	$P_4 = 0.0625$

number of samples (i.e., N). This means a considerably large number of samples will be needed in order to approximate unreliability if the network is highly reliable. In this case, the system failure event is referred to as a *rare event*. It is rare because it occurs once in large number of samples (say once every 10^6 samples!).

Several variance-reduction techniques are used to remedy the situation. Stratified sampling is used in Listing 12.3. Antithetic sampling is used in Listing 12.4. The new code is on lines 22-27. Finally, dagger sampling is used in Listing 12.5.

Listing 12.2
Computing unreliability for the graph in Figure 12.2 using crude Monte Carlo simulation.

```python
1  from random import *
2  from statistics import *
3
4  q = 0.3        # Prob. of link failure
5  N = 100000     # Number of trials
6  L = 4          # Number of links
7
8  # Check if network is connected
9  def Phi(s):
10     if s[0] == s[1] == 0 or s[0] == s[2] == 0 or \
11        s[1] == s[3] == 0 or s[2] == s[3] == 0:
12         return 1
13     else:
14         return 0
15
16  # Crude Monte Carlo simulation
17  rv = []         # Realization of a Bernoulli random variable
18  for i in range(N):
19      s = [0]*L
20      for j in range(L):
21          if random() > q:
22              s[j] = 1
23      rv.append(Phi(s))
24
25  # Result
26  print("Unreliability = ", round(mean(rv), 4)) # 0.2593
27  print("Variance = ", round(variance(rv), 4)) # 0.1921
```

Listing 12.3
Computing unreliability for the graph in Figure 12.2 using stratified sampling.

```python
from random import *

from math import *

from statistics import *


q = 0.5          # Prob. of link failure

N = 100000       # Number of trials

L = 4            # Number of links


K = L   # Number of strata

P = [0.0625, 0.25, 0.375, 0.25, 0.0625]  # Pi for each
    stratum i


# Number of samples from each stratum

N_i = [int(p * N) for p in P]


# Generate a sample

# n = Number of UP links

def samp(n):

    if n == 0:

        return [0, 0, 0, 0]

    elif n == 4:

        return [1, 1, 1, 1]

    elif n == 1:

        i = randint(0, 3)

        s = [0] * L

        s[i] = 1

        return s

    elif n == 2:

        idx = sample([0, 1, 2, 3], 2)    # Unique indexes

        s = [0] * L

        s[ idx[0] ] = 1

        s[ idx[1] ] = 1
```

```
32          return s
33      elif n == 3:
34          idx = sample([0, 1, 2, 3], 3)
35          s = [0] * L
36          s[ idx[0] ] = 1
37          s[ idx[1] ] = 1
38          s[ idx[2] ] = 1
39          return s

40
41  # Check if network is connected
42  def Phi(s):
43      if s[0] == s[1] == 0 or s[0] == s[2] == 0 or \
44          s[1] == s[3] == 0 or s[2] == s[3] == 0:
45          return 1
46      else:
47          return 0

48
49  rv = []
50  for i in range(K+1):
51      m = N_i[i]
52      for j in range(m):
53          s = samp(i)
54          rv.append( Phi(s) )

55
56  # Result
57  print("Unreliability = ", round(mean(rv), 4)) # 0.5636
58  print("Variance = ", round(variance(rv), 4)) # 0.246
```

Listing 12.4
Computing unreliability for the graph in Figure 12.2 using antithetic sampling.

```
1  from random import *
```

```python
from statistics import *

q = 0.3          # Prob. of link failure
N = 100000       # Number of trials
L = 4            # Number of links


# Check if network is connected
def Phi(s):
    if s[0] == s[1] == 0 or s[0] == 0 and s[2] == 0 or \
        s[1] == s[3] == 0 or s[2] == s[3] == 0:
        return 1
    else:
        return 0


# Antithetic Monte Carlo simulation
rv = []
for i in range(N):
    s1 = [0]*L
    s2 = [0]*L
    for j in range(L):
        u = random()
        if u > q:        s1[j] = 1
        if (1 - u) > q: s2[j] = 1

    val = (Phi(s1) + Phi(s2) ) / 2
    rv.append(val)


# Result
print("Unreliability = ", round(mean(rv), 4)) # 0.2597
print("Variance = ", round(variance(rv), 4)) # 0.0784
```

Listing 12.5
Computing unreliability for the graph in Figure 12.2 using dagger sampling.
The number of samples is significantly less.

```
1   from random import *
2   from statistics import *
3   from math import *
4
5   q = 0.3          # Prob. of link failure
6   N = 30000        # Number of trials
7   L = 4            # Number of links
8
9   # Check if network is connected
10  def Phi(s):
11      if s[0] == s[1] == 0 or s[0] == 0 and s[2] == 0 or \
12          s[1] == s[3] == 0 or s[2] == s[3] == 0:
13          return 1
14      else:
15          return 0
16
17  # Antithetic Monte Carlo simulation
18  rv = []
19  for i in range(N):
20      s1 = [1]*L
21      s2 = [1]*L
22      s3 = [1]*L
23      for j in range(L):
24          u = random()
25          if 0 < u <= q:
26              s1[j] = 0
27          elif q < u <= 2*q:
28              s2[j] = 0
29          elif 2*q < u <= 3*q:
30              s3[j] = 0
31
```

```
32      rv.append(Phi(s1))

33      rv.append(Phi(s2))

34      rv.append(Phi(s3))

35

36  # Result

37  print("Unreliability = ", round(mean(rv), 4))  # 0.2617

38  print("Variance = ", round(variance(rv), 4))  # 0.1932
```

12.2 PACKET DELIVERY OVER A WIRELESS CHANNEL

Consider a transmitter that sends packets to a receiver over a point-to-point wireless channel as shown in Figure 12.3. The transmitter has a buffer which can hold up to B packets. For each packet, the transmitter has T transmission attempts. If they all fail, the packet is dropped. A transmission may fail because of the time-varying nature of the wireless channel. The probability that a packet transmission is unsuccessful is denoted by P_{err}. Hence, the probability of a successful transmission is $1 - P_{err}$.

Upon its arrival, a packet will be lost if the queue is full. This situation is captured by the Loss event and the condition on the edge connecting it with the Arrival event. If the packet, however, enters the queue, it will be scheduled for transmission once it is at the head of the queue. For each Transmit event, two events are scheduled: Receive and Timeout. If a packet is successfully received, its corresponding Timeout event is removed from the event list and the next packet in the queue is scheduled for transmission.

On the other hand, if a packet is not delivered to the receiver, its Timeout event will eventually fire and cause a re-transmission if the number of trans-

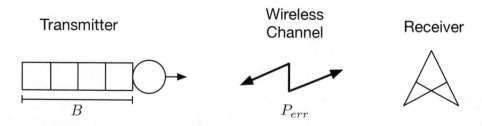

Figure 12.3

A point-to-point wireless system. The transmitter has a buffer which can store up to B packets. The probability that a transmission attempt is successful is $1 - P_{err}$.

Table 12.3
State variables of the event graph in Figure 12.4.

State Variable	Description
Q	Number of packets in the queue
C	State of the wireless channel ("0": Available; "1": Occupied)
L	Number of lost packets
D	Number of dropped packets
T	Number of transmission attempts

mission attempts is still below the preset threshold (i.e., $T < \tau$). Otherwise, a Drop event is scheduled. After the current packet is discarded, the next packet in the queue is scheduled for transmission. Table 12.3 shows the state variables used in the construction of the event graph for this system. The event graph is shown in Figure 12.4. There are two system parameters which are the maximum allowed number of transmission attempts for every packet (τ) and the packet error rate of the wireless channel (P_{err}).

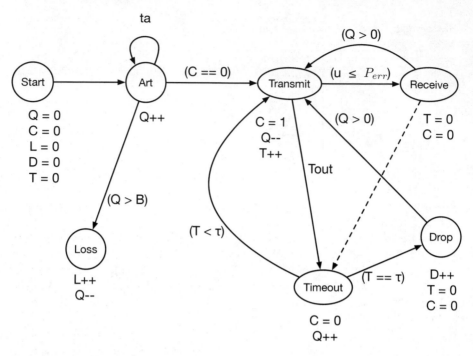

Figure 12.4
Event graph for the system in Figure 12.3.

```
Listing 12.6
Python implementation of the event graph in Figure 12.4

1   from random import *
2   from bisect import *
3   from statistics import *
4
5   # Simulation parameters
6   n = 1000   # Number of packets to be simulated
7   lamda = 0.7
8   P_err = 0.99
9   tau = 3
10  Tout = 1   # Length of timeout period
11  B = 10   # Size of transmitter buffer
12
13  # Initialization
```

```
14  clock = 0.0
15  evList = []
16  count = 0  # Used for counting simulated packets and as
         Pkt_ID
17  evID = 0  # Unique ID for each event
18  Timeout_Event = None  # Reference to currently pending
       timeout event
19
20  # Insert an event into the event list
21  def insert(ev):
22      insort_right(evList, ev)
23
24  # Remove an event from the event list
25  def cancel(ev):
26      evList.remove(ev)
27
28  # Initialize state variables
29  Q = 0
30  C = 0
31  L = 0
32  D = 0
33  T = 0
34
35  # Output variables
36  Num_Received_Pkts = 0  # Pkts received successfully
37  Arr_Time = [0] * n
38  Rec_Time = [0] * n
39
40  # Event generators
41  def Gen_Arr_Evt(clock):
42      global count, n, lamda, evID
43      if count < n:
44          insert( (clock + expovariate(lamda), evID, count,
         Handle_Arr_Evt) )
```

```
45          count += 1
46          evID += 1
47
48  def Gen_Loss_Evt(clock, Pkt_ID):
49      global evID
50      evID += 1
51      insert( (clock, evID, Pkt_ID, Handle_Loss_Evt) )
52
53  def Gen_Transmit_Evt(clock, Pkt_ID):
54      global evID
55      evID += 1
56      insert( (clock, evID, Pkt_ID, Handle_Transmit_Evt) )
57
58  def Gen_Receive_Evt(clock, Pkt_ID):
59      global evID
60      evID += 1
61      insert( (clock, evID, Pkt_ID, Handle_Receive_Evt) )
62
63  def Gen_Drop_Evt(clock, Pkt_ID):
64      global evID
65      evID += 1
66      insert( (clock, evID, Pkt_ID, Handle_Drop_Evt) )
67
68  def Gen_Timeout_Evt(clock, Pkt_ID):
69      global Timeout_Event, evID
70      evID += 1
71      Timeout_Event = (clock + Tout, evID, Pkt_ID,
        Handle_Timeout_Evt)
72      insert( Timeout_Event )
73
74  # Event handlers
75
76  def Handle_Arr_Evt(clock, Pkt_ID):
77      global Q, lamda
```

```
78      Q += 1
79      Gen_Arr_Evt(clock + expovariate(lamda))
80      if C == 0:
81          Gen_Transmit_Evt(clock, Pkt_ID)
82      if Q > B:
83          Gen_Loss_Evt(clock, Pkt_ID)
84      # Output variable
85      Arr_Time[Pkt_ID] = clock
86
87  def Handle_Loss_Evt(clock, Pkt_ID):
88      global Q, L
89      L += 1
90      Q -= 1
91
92  def Handle_Transmit_Evt(clock, Pkt_ID):
93      global C, Q, T, P_err
94      C = 1
95      Q -= 1
96      T += 1
97      Gen_Timeout_Evt(clock, Pkt_ID)
98      if random() <= (1 - P_err):
99          Gen_Receive_Evt(clock, Pkt_ID)
100
101 def Handle_Receive_Evt(clock, Pkt_ID):
102     global C, T, Q, Num_Received_Pkts
103     C = 0
104     T = 0
105     cancel(Timeout_Event)
106     if Q > 0:
107         Gen_Transmit_Evt(clock, Pkt_ID + 1)   # Next packet
        in queue
108     # Output variable
109     Num_Received_Pkts += 1
110     Rec_Time[Pkt_ID] = clock
```

```python
112  def Handle_Drop_Evt(clock, Pkt_ID):
113      global D, T, C, Q
114      C = 0
115      T = 0
116      D += 1
117      Q -= 1
118      if Q > 0:
119          Gen_Transmit_Evt(clock, Pkt_ID + 1)
120
121  def Handle_Timeout_Evt(clock, Pkt_ID):
122      global T, C, Q
123      C = 0
124      Q += 1
125      if T == tau:
126          Gen_Drop_Evt(clock, Pkt_ID)
127      elif T < tau:
128          Gen_Transmit_Evt(clock, Pkt_ID)  # Re-transmit same
         packet
129
130  # Generate initial events
131  Gen_Arr_Evt(0.0)
132
133  # Simulation loop
134  while evList:
135      ev = evList.pop(0)
136      clock = ev[0]
137      Pkt_ID = ev[2]
138      ev[3](clock, Pkt_ID) # call event handler
139
140  # Statistical summary
141  Delay = []
142  for i in range(n):
143      if Rec_Time[i] > 0:
```

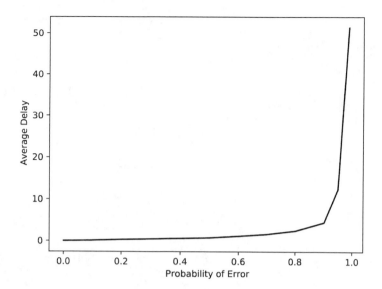

Figure 12.5
Average packet delay increases as the quality of the wireless channel degrades.

```
144        Delay.append( Rec_Time[i] - Arr_Time[i] )
145   print("Average delay through the system = ", round(mean(
          Delay), 2))
146   print("Percentage of received packets = ", round((
          Num_Received_Pkts / n) * 100, 1))
```

Two measures of performance are considered: average delay and percentage of successfully received packets. They both depend on the quality of the wireless channel. Figure 12.5 shows that the average delay increases as the quality of the wireless channel degrades. Similarly, as shown in Figure 12.6, the percentage of received packets significantly decreases as the error rate of the wireless channel increases.

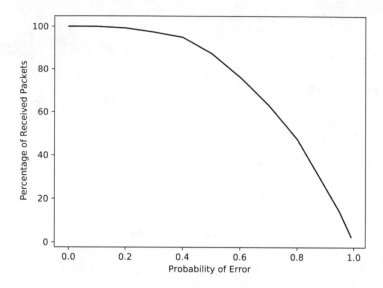

Figure 12.6
Percentage of delivered packets drops as the quality of the wireless channel degrades.

12.3 SIMPLE ARQ PROTOCOL

Automatic Repeat Request (ARQ) is an error-control technique used in computer networks. It is based on the use of *acknowledgment* messages and *timeout* interrupts. Basically, when a message is transmitted through a communication channel, it may not arrive to the receiver because it is either lost or it is in error. The receiver cannot interpret an erroneous packet. In this case, there are two scenarios (see Figure 12.7):

1. The receiver receives the packet and the packet is not in error. In this case, the receiver has to notify the sender; i.e., it acknowledges the reception of the packet.

2. The receiver does not receive the packet. As a result, it cannot send back an acknowledgment since it has not received the packet. Hence, it is the responsibility of the sender to detect this event by using timeout. A *timeout* is a pre-defined period of time during which an acknowledgment message is expected. If this timer expires, then the packet must be retransmitted.

The event graph in Figure 12.8 is for a simple stop-and-wait ARQ protocol between a sender and receiver connected by a wireless channel. The wireless channel is fully characterized by its Packet Error Rate PER, which is the probability that a packet is lost or corrupted while in transit through the channel. The probability of a successful reception is thus $1 - PER$.

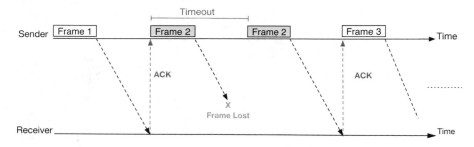

Figure 12.7
Behavior of the simple stop-and-wait ARQ protocol with two possibilities: acknowledgment and frame loss.

Assuming there is at least one packet in the transmission buffer of the sender, the initial event is a transmission event. This is just a convention to simplify the example. A transmission event (`Transmit`) schedules two events: `Receive` and `Timeout`. The successful reception of a packet is simulated by generating a uniform random number and comparing it against the probability $1 - PER$. If the condition is satisfied, the `Receive` event occurs after a period of time equal to the total reception time t_{rec}. This time accounts for both the packet transmission time and propagation delay through the channel.

Upon the reception of a packet, the `Receive` event schedules an ACK event, which immediately schedules the next transmission event. Of course, the pending `Timeout` event for the just received packet must be cancelled by the ACK event. The `Timeout` event, however, occurs if there is no `Receive` event scheduled for the current packet. The main purpose of this event is to trigger a re-transmission of the current packet after a period of time equal to t_{out}.

Listing 12.7 gives the code for implementing the event graph in Figure 12.8. Note that simulation parameters are contained in a Python dictionary, which is passed to every event handler. This is just another way to access global parameters in your simulation program.

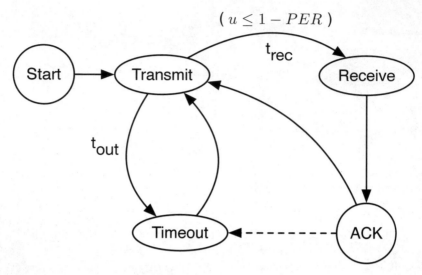

Figure 12.8
Event graph for the simple stop-and-wait ARQ protocol.

```
Listing 12.7
Python implementation of the event graph of the simple stop-and-wait ARQ
protocol in Figure 12.8.

 1  import random as rnd
 2  import queue
 3  import statistics as stat
 4
 5  # Define a dictionary to hold the simulation parameters
 6  param = {'Timeout_Duration': 1,
 7      'P' : 0.2, # Packet Error Rate (PER)
 8      'Frame_Trans_Time': 1, # Frame transmission time
 9      'Num_Frames': 10000
10      }
11
12  #--------------- Global Variables ---------------
13  Frames_Received = 0.0
14  Count_Frames = 0.0
15  clock = 0.0
16  evList = queue.PriorityQueue()
```

```python
17
18  # Unique ID for each event
19  evID = 0
20
21  #-------------- Event Generators --------------
22  # REG for the sender start event
23  def sender_start_event (clock, param):
24    global evID
25    ev = (clock, evID, sender_start_event_handler)
26    evID += 1
27    return ev
28
29  # REG for the receiver start event
30  def receiver_start_event (clock, param):
31    global evID
32    ev = (clock, evID, receiver_start_event_handler)
33    evID += 1
34    return ev
35
36  # REG for the frame transmission event
37  def frame_trans_event (clock, param):
38    global evID, Count_Frames
39    if(Count_Frames < param['Num_Frames']):
40      Count_Frames += 1
41      ev = (clock, evID, frame_trans_event_handler)
42      evID += 1
43      return ev
44
45  # REG for the timeout event
46  def timeout_event (clock, param):
47    global evID
48    t = param['Timeout_Duration']
49    ev = (clock+t, evID, timeout_event_handler)
50    evID += 1
```

```
51    return ev
52
53  # REG for the frame reception event
54  def frame_reception_event (clock, param):
55    global evID
56    t = param['Frame_Trans_Time']
57    ev = (clock+t, evID, frame_reception_event_handler)
58    evID += 1
59    return ev
60
61  # REG for the acknowledgment event
62  def ack_event (clock, param):
63    global evID
64    ev = (clock, evID, ack_reception_event_handler)
65    evID += 1
66    return ev
67
68  #--------------- Event Handlers ---------------
69  # Event handler for the sender start event
70  def sender_start_event_handler (clock, param):
71    global Count_Frames
72    Count_Frames = 0.0
73    # Schedule the first frame transmission event
74    schedule_event( frame_trans_event (clock, param) )
75
76  # Event handler for the receiver start event
77  def receiver_start_event_handler (clock, param):
78    global Frames_Received
79    Frames_Received = 0.0
80
81  # Event handler for the frame transmission event
82  def frame_trans_event_handler (clock, param):
83    # Generate a frame reception event if frame is going
84    # to be successfully received
```

```python
85      if rnd.random() <= param['P']:
86        # Frame is damaged. Generate a timeout event
87        schedule_event( timeout_event (clock, param) )
88      else:
89        # Frame is successfully delivered
90        schedule_event(
91                frame_reception_event (clock, param) )
92
93  # Event handler for the frame reception event
94  def frame_reception_event_handler (clock, param):
95    global Frames_Received
96    Frames_Received += 1
97    schedule_event( ack_event (clock, param) )
98
99  # Event handler for the ack event
100 def ack_reception_event_handler (clock, param):
101   schedule_event( frame_trans_event (clock, param) )
102
103 # Event handler for the timeout event
104 def timeout_event_handler (clock, param):
105   global Count_Frames
106   # Re-transmit the frame again
107   Count_Frames = Count_Frames - 1
108   schedule_event( frame_trans_event (clock, param) )
109
110 # Insert an event into the event list
111 def schedule_event (ev):
112   global evList
113   if ev != None:
114     evList.put (ev)
115
116 #------ Start Simulation ------
117
118 # 1. Initialize sender and receiver
```

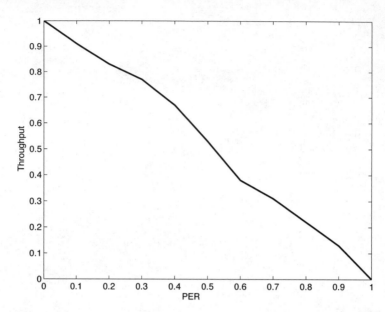

Figure 12.9
Throughput deteriorates as the packet error rate increases.

```
119   schedule_event( sender_start_event (clock, param) )
120   schedule_event( receiver_start_event (clock, param) )
121
122   # 2. Run the simulation loop
123   while not evList.empty():
124     ev = evList.get()
125     clock = ev[0]
126     ev[2](clock, param)
```

An interesting measure of performance is how throughput changes with the change in packet error rate (PER). Figure 12.9 shows that throughput deteriorates as PER increases. This is expected since a bad wireless channel will significantly reduce the number of successfully received packets. The code is self-explanatory and the reader is encouraged to identify the event generator and handler for each event type in the event graph.

12.4 SUMMARY

Moving from a system description to a simulation program is not a trivial task. A simulation model must be constructed using event graphs before any code can be written. Then, the simulation model can be translated into code using the concepts and conventions discussed in Chapter 7. The purpose of this chapter was to reinforce this skill.

12.5 EXERCISES

12.1 Study the relationship between the average packet delay and transmission attempt threshold by extending the program in Listing 12.6.

12.2 Identify a redundant event in the event graph in Figure 12.8.

   a. Re-draw the event graph after removing the redundant event.

   b. Is the new event graph equivalent to the original one?

Overview of Python

"I was looking for a hobby. So, I decided to develop a new computer language."
−Guido van Rossum

This appendix serves as an introduction to the Python programming language. It covers all the language features used in developing the examples in the book. For more information about Python, you should consult one of the many books on the Python programming language.

A.1 BASICS

Python is an interpreted language. When you type *python* at the command prompt, the Python prompt ($>>>$) appears where you can start typing Python statements. Listing A.1.1 shows how a new Python interactive session cab be started.

Listing A.1.1
Starting a new Python interactive session.

```
1  C:/> python
2  Python 3.6.2 .... more information will be shown
3  >>> 1 + 1
4  2
5  >>>
```

You can store your code in a file and call the Python interpreter on the file from the command prompt. The command prompt of the operating system will appear after the execution of the file finishes. Listing A.1.2 is an example of running a Python file containing a program that adds two numbers.

Listing A.1.2
Running a Python program from the command line.

```
1  C:/> python my_prog.py
2  Enter the two numbers to add: 1 3
3  Result = 4
4  C:/>
```

In Python, variables are not explicitly created. A variable is created when its name is used for the first time on the left-hand side of an assignment statement. The logical operators are *and*, *or*, and *not*.

Listing A.1.3 shows a simple Python program (or script). There is no *main* function. When the Python interpreter is called on this file, execution starts from the top of the file. The program is executed line by line. The execution of the file stops when its end is reached.

Not like C and Java, a semi-colon is not used to indicate the end of a statement. Instead, Python relies heavily on the use of white spaces to indicate where a statement starts and ends. For example, the new-line character signifies the end of a statement.

A block of code starts after a line ending with a colon. For example, see the for-loop block in Listing A.1.3. A statement belongs to the body of the for-loop as long as it is indented to the right. The statements making up the body of the for-loop must also be aligned. This is how the Python interpreter can find the start and end of the for-loop.

There are two levels in the source file in Listing A.1.3. The first level is for every statement in the file. The second level is only for the statements in the body of the for-loop. If an *if-statement* is to be used inside the for-loop, a third level will be introduced to indicate the body of the if-statement. A new level can be introduced by simply pressing on the *Tab* key on the keyboard.

Listing A.1.3
A Python source file. It can also be referred to as a Python script.

```
1  from random import choice
2
3  lower = input("Enter smallest number: ")
4  upper = input("Enter largest number: ")
```

```
5   n = input("How many numbers do you want to generate? ")
6
7   # Parse strings into integers
8   lower = int(lower)
9   upper = int(upper)
10  n = int(n)
11
12  # Construct a list from a range object
13  numbers = list(range(lower, upper + 1))
14
15  selection = []
16
17  for i in range(n):
18      r = choice(numbers)
19      selection.append(r)
20      numbers.remove(r)
21
22  print( "Your numbers: ", selection )
```

A.2   INPUT AND OUTPUT

Listing A.2.1 shows two functions for reading input from the user and printing output to the console.

Listing A.2.1
Input and output functions.

```
1   >>> m = input("Enter the mean service time: ")
2   Enter the mean service time: 5      # Enter number 5
3   >>> m
4   5
5   >>> print( "You entered: ", m )
6   You entered:   5
```

A.3 BITWISE OPERATORS

Listing A.3.1
Binary operations on integer numbers.

```python
a = 10   # 0000 1010
b = 25   # 0001 1001

# AND
c = a & b
print(c)   # 0000 1000 (8)

# OR
c = a | b
print(c) # 0001 1011 (27)

# XOR
c = a ^ b
print(c) # 0001 0011 (19)

# Ones Complement
# Numbers are in 2's complement representation
c = ~a
print(c) # 1111 0101 (-11)

# Right Shift
c = a >> 2
print(c) # 0000 0010 (2)

# Left Shift
c = a << 2
print(c) # 0010 1000 (40)
```

Listing A.5.1
Transposing a matrix using the zip function. Matrix is first unpacked using the start (*) operator.

```
1  matrix = [ [1, 2], [3, 4] ]
2  matrix_transposed = list(zip( *matrix ))
3  # *matrix => [1, 2] [3, 4]
4  print(matrix_transposed) # [(1, 3), (2, 4)]
```

A.6 GENERATING RANDOM NUMBERS AND RANDOM VARIATES

The *random* module provides RNGs and RVGs for various probability distributions. You need to first include the module in your Python script by importing it. Listing A.6.1 shows the procedure.

Listing A.6.1
Importing the *random* module and calling some of the functions inside it.

```
1  >>> import random
2  >>> random.random() # Returns a floating-point number in
3  0.8545672259166788  # the range (0,1)
4
5  >>> random.randrange(1,6)  # Returns an integer in the
6  4                          # range [1, 6)
7
8  >>> random.uniform(1, 3)   # Returns a floating-point number
9  1.290486289287417          # in the range [1, 3)
10
11 >>> random.normalvariate(1,0) # Returns a normal variate where
12 1.0                           # mean = 1 and stdDev = 0
13
14 >>> random.expovariate(3)     # Returns an exponential variate
15 0.06953873605855697           # with mean 1/3
```

```
16
17  >>> random.choice([1,2,3,4,5,6]) # Returns a random element
        from
18  5                                  # the input sequence
19
20  >>> random.sample([1,2,3,4,5,6], 3) # Randomly choose three
21  [6, 1, 2]                             # elements from the
        given
22                                        # sequence
```

A.7   IMPLEMENTING THE EVENT LIST

A.7.1   Priority Queue

Listing A.7.1
Implementing the event list using the *queue* module.

```
1  import queue
2  from queue import Queue
3
4
5  Event_List = queue.PriorityQueue()
6
7  for item in ((10, "Arrival"), (5, "Departure"), (2, "
       Fully_Charged")):
8      Event_List.put(item)
9
10  while not Event_List.empty():
11      print(Event_List.get())
```

A.7.2   Heap Queue

Listing A.7.2
Implementing the event list using the *hqueue* module.

```
1   import heapq
2   from heapq import *
3
4   Event_List =[]
5
6   heappush(Event_List, (10, "Arrival"))
7   heappush(Event_List, (5, "Departure"))
8   heappush(Event_List, (2, "Fully_Charged"))
9
10  # Print the first item in the heap
11  print ( heappop(Event_List) )
```

A.7.3   Sorting a List

Listing A.7.3
Implementing the event list by sorting a list.

```
1   # The first field is always the time
2   e1 = (10, "Arrival")
3   e2 = (5, "Departure")
4   e3 = (2, "Fully_Charged")
5
6   Event_List = []
7
8   Event_List += [e1]
9   Event_List += [e2]
10  Event_List += [e3]
11  Event_List.sort()
12
13  print(Event_List)
```

A.8 PASSING A FUNCTION NAME AS AN ARGUMENT

> Listing A.8.1
> The name of the function can be stored in a list and then used to call the function.

```
1  def add():
2      print ( "Add" )
3
4  def sub():
5      print ( "Sub" )
6
7  a = [add, sub]
8
9  for i in range(len(a)):
10     a [i] ( )      # Add two parentheses and include arguments
       ,
11                    # if any
```

> Listing A.8.2
> The name of the function can be passed as an argument to another function.

```
1  def doIt (func, x, y):
2      z = func (x, y)
3      return z
4
5  def add (arg1, arg2):
6      return arg1 + arg2
7
8  def sub (arg1, arg2):
9      return arg1 - arg2
10
11 print ("Addition:")
```

```
12  print ( doIt (add, 2, 3) )       # Passing the name of the
        function
13                                    # and its arguments
14
15  print ("Subtraction:")
16  print ( doIt (sub, 2, 3) )
```

A.9 TUPLES AS RECORDS

Listing A.9.1
A *tuple* can be used as a record that represents an item in the event list.

```
1   def Handle_Event_1():
2       print ( "Event_1" )
3
4   def Handle_Event_2():
5       print ( "Event_2" )
6
7   Event_List = [(1.3, Handle_Event_1), (3.3, Handle_Event_2),
8                                         (4.5, Handle_Event_1)]
9
10  for ev in Event_List:
11      (time , event_handler) = ev
12      event_handler ( )       # Add two parentheses and include
13                              # arguments, if any
```

A.10 PLOTTING

Listing A.10.1
Code for generating Figure 4.12(b).

```
1   from random import *
2   from math import *
```

```python
3   from matplotlib.pyplot import *
4   from numpy import *
5
6   def pdf(x, a, b, c):
7       if x < a:
8           return 0
9       elif x >= a and x < c:
10          return (2 * (x-a)) / ((b-a)*(c-a))
11      elif x == c:
12          return 2 / (b-a)
13      elif x > c and x <= b:
14          return (2 * (b-x)) / ((b-a)*(b-c))
15      elif x > b:
16          return 0
17      else:
18          print("Error")
19
20
21  a = 1
22  b = 10
23  c = 7
24
25  X = arange(0, b+1, 0.1)
26  Y = []
27
28  xlabel("X", fontsize=15)
29  ylabel("f(x)", fontsize=15)
30
31  gca().axes.get_xaxis().set_ticks( np.arange(0, b+1, 1.0) )
32
33  for x in X:
34      Y.append( pdf(x, a, b, c) )
35
36  plot(X, Y, linewidth=2)
```

```
37
38  # Show figure on screen
39  show()
40
41  # Save figure to hard disk
42  savefig("triangular_pdf.pdf", format="pdf", bbox_inches="
        tight")
```

Listing A.10.2
Code for generating Figure 10.6(a).

```
1   from random import *
2   from math import *
3   from matplotlib.pyplot import *
4   from numpy import *
5
6   def pdf(x):
7       k = 10
8       theta = 1.0
9       return (x**(k-1) * theta**k * exp(-1 * theta * x)) /
            factorial(k-1)
10
11  X = arange(0, 50, 0.1)
12  Y = []
13  for x in X:
14      Y.append( pdf(x) )
15
16  xlabel("Y")
17  ylabel("P(y)")
18
19  # Hide numbers along y-axis
20  gca().axes.get_yaxis().set_ticklabels([])
```

```
21  # Remove ticks along y-axis
22  gca().axes.yaxis.set_tick_params(width=0)
23
24  plot(X, Y, linewidth=2)
25  savefig("erlang_plot_pdf.pdf", format="pdf", bbox_inches="
        tight")
26
27  # Compute the mean
28  mean = 0
29  for i in range( len(X) ):
30      mean = mean + X[i] * Y[i]
31
32  print("Mean = ", mean)
```

Listing A.10.3
Code for generating Figure 10.6(b).

```
1   from random import *
2   from math import *
3   from matplotlib.pyplot import *
4   from statistics import *
5
6   def Erlang():
7       k = 10
8       theta = 1.0
9       y = 0
10      for i in range(k):
11          u = random()
12          x = (-1 / theta) * log(u)      # Exponential variate
13          y = y + x
14
15      return y
```

```
16
17  N = 100000
18  v = []
19  for i in range(N):
20      v.append( Erlang() )
21
22  bins = 100
23
24  w = [1 / len(v)] * len(v)
25
26  hist(v, bins, weights = w)
27
28  xlabel("Y")
29  ylabel("P(y)")
30
31  # Hide numbers along y-axis
32  gca().axes.get_yaxis().set_ticklabels([])
33  # Remove ticks along y-axis
34  gca().axes.yaxis.set_tick_params(width=0)
35
36  savefig("erlang_plot_hist.pdf", format="pdf", bbox_inches="
        tight")
37
38  print("Mean = ", mean(v))
```

An Object-Oriented Simulation Framework

"Object-oriented programming is an exceptionally bad idea."
—Edsger Dijkstra

The object-oriented paradigm is one of the most widely used programming paradigms. This is because of its many benefits like code organization and reuse. This appendix contains the details of a simulation framework that use the object-oriented features of Python. This framework is inspired by the Java-based framework described in [2].

Listing B.1
Event.

```python
class Event:
    def __init__(self, _src, _target, _type, _time):
        self.src = _src
        self.target = _target
        self.type = _type
        self.time = _time

    def __eq__(self, other):
        return self.__dict__ == other.__dict__
```

Listing B.2
Simulation Entity.

```python
from event import Event

class SimEntity:

    def __init__(self, _scheduler, _id):
        self.scheduler = _scheduler
        self.id = _id

    def schedule(self, target, type, time):
        ev = Event(self, target, type, time)
        self.scheduler.insert(ev)

    def cancel(self, ev):
        self.scheduler.remove(ev)

    def evHandler(self, ev):
        pass
```

Listing B.3
Event list and scheduler.

```python
# The variable self.count_events counts events generated.
    The number of events executed will also be equal to this
# number.
#
# You insert event based on its time. If there are two
    events occurring at the same time, the second sorting
# criterion is to use the id of the target. The third
    sorting criterion is to use the id of the source.

from queue import PriorityQueue
```

```python
8
9   class Scheduler:
10
11      def __init__(self, _Max_Num_Events):
12          self.evList = PriorityQueue()
13          self.time = 0.0
14          self.count_events = 0
15          self.Max_Num_Events = _Max_Num_Events
16
17      def insert(self, ev):
18          if ( self.count_events < self.Max_Num_Events ):
19              self.count_events = self.count_events + 1
20              self.evList.put( (ev.time, self.count_events, ev
    ) )
21
22      def remove(self, ev):
23          _evList = PriorityQueue()
24          for i in range(self.evList.qsize()):
25              tmp = self.evList.get()
26              _ev = tmp[2]
27              if not _ev == ev:
28                  _evList.put(tmp)
29          self.evList = _evList
30
31      def head(self):
32          ev = self.evList.get()
33          self.time = ev[2].time
34          return ev[2]
35
36      def run(self):
37          count = 0
38          while( not self.empty() ):
39              ev = self.head()
40              self.time = ev.time
```

```
41          count += 1
42          ev.target.evHandler(ev)
43
44    def empty(self):
45        return self.evList.empty()
46
47    def reset(self):
48        self.evList = None
49        self.time = 0.0
50        self.count_events = 0
```

Listing B.4
Example 1.

```
1  from scheduler import Scheduler
2  from simEntity import SimEntity
3
4
5  class Node(SimEntity):
6      def __init__(self, _scheduler, _id):
7          super(Node, self).__init__(_scheduler, _id)
8          self.schedule(self, "Self_Message", self.scheduler.
   time + 2.0)
9
10     def evHandler(self, ev):
11         print( ev.type + " From " + str(ev.src.id) + " To "
   + str(ev.target.id) + " @ " + str(ev.time) )
12         self.schedule(self, "Hi", self.scheduler.time + 2.0)
13
14
15 scheduler = Scheduler(3)
16
```

```
17  Node(scheduler, 1)

18

19  scheduler.run()
```

Listing B.5
Example 2.

```
1  from scheduler import Scheduler
2  from simEntity import SimEntity
3
4  class Node(SimEntity):
5
6      def __init__(self, _scheduler, _id):
7          super(Node, self).__init__(_scheduler)
8          self.id = _id
9          self.schedule(self, "Initialize", self.scheduler.
    time + 2.0)
10
11     def setNeighbor(self, n):
12         self.neighbor = n
13
14     def evHandler(self, ev):
15         print( ev.type + " From " + str(ev.src.id) + " To "
    + str(ev.target.id) + " @ " + str(ev.time) )
16         self.schedule(self.neighbor, "Hi", self.scheduler.
    time + 3.0)
17
18
19
20
21  scheduler = Scheduler(4)
22
```

```
23  n1 = Node(scheduler, 1)
24  n2 = Node(scheduler, 2)
25
26  n1.setNeighbor(n2)
27  n2.setNeighbor(n1)
28
29  scheduler.run()
```

Listing B.6
Example 3.

```
1   # Remove n1.setNeighbor(n2) && n2.setNeighbor(n1)
2   # Use a link to connect the two nodes
3   # A link has two ends: a & b
4
5   from scheduler import Scheduler
6   from simEntity import SimEntity
7
8   class Node(SimEntity):
9
10      def __init__(self, _scheduler, _id):
11          super(Node, self).__init__(_scheduler, _id)
12          self.schedule( self, "Initialize", self.scheduler.
        time + 2.0 )
13
14      def setNeighbor(self, n):
15          self.neighbor = n
16
17      def evHandler(self, ev):
18          print( ev.type + " From " + str(ev.src.id) + " To "
        + str(ev.target.id) + " @ " + str(ev.time) )
```

```python
19          self.schedule( self.neighbor, "Hi", self.scheduler.
       time + 3.0 )
20
21
22  class Link(SimEntity):
23
24      def __init__(self, _scheduler, _id):
25          super(Link, self).__init__(_scheduler, _id)
26
27      def setNeighbors(self, _a, _b):
28          self.a = _a
29          self.b = _b
30
31      def evHandler(self, ev):
32          print( ev.type + " From " + str(ev.src.id) + " To "
       + str(ev.target.id) + " @ " + str(ev.time) )
33          if( ev.src.id == self.a.id ):
34              self.schedule( self.b, "Hi", self.scheduler.time
           + 3.0 )
35          else:
36              self.schedule( self.a, "Hi", self.scheduler.time
           + 3.0 )
37
38
39
40  scheduler = Scheduler(6)
41
42  n1 = Node(scheduler, 1)
43  n2 = Node(scheduler, 2)
44  l = Link(scheduler, 3)
45
46  n1.setNeighbor(l)
47  n2.setNeighbor(l)
48  l.setNeighbors(n1, n2)
```

```
49
50    scheduler.run()
```

Listing B.7
Example 4.

```
1    from scheduler import Scheduler
2    from simEntity import SimEntity
3    from event import Event
4
5    class Node(SimEntity):
6        def __init__(self, _scheduler, _id):
7            super(Node, self).__init__(_scheduler, _id)
8            self.schedule(self, "Self_Message", self.scheduler.
     time + 5.0)
9            self.schedule(self, "Self_Message", self.scheduler.
     time + 3.0)
10           self.schedule(self, "Self_Message", self.scheduler.
     time + 4.0)
11           self.schedule(self, "Self_Message", self.scheduler.
     time + 1.0)
12           self.schedule(self, "Self_Message", self.scheduler.
     time + 2.0)
13           ev = Event(self, self, "Self_Message", self.
     scheduler.time + 1.0)
14           self.cancel(ev)
15
16       def evHandler(self, ev):
17           print( ev.type + " From " + str(ev.src.id) + " To "
     + str(ev.target.id) + " @ " + str(ev.time) )
18
19
```

```
20
21   scheduler = Scheduler(5)
22
23   Node(scheduler, 1)
24
25   scheduler.run()
```

Listing B.8
M/M/1.

```
1    # IAT = Average Inter-Arrival Time
2    # ST = Average Service Time
3    # Size of packet is its service time (in time units not bits
         )
4    # Station contains a queue (Q) and server (S)
5
6    from scheduler import Scheduler
7    from simEntity import SimEntity
8    import random as rnd
9    import queue
10
11   class TrafficGen(SimEntity):
12
13       def __init__(self, _scheduler, _station, _id, _IAT =
         1.0, _ST = 1.0):
14           super(TrafficGen, self).__init__(_scheduler, _id)
15           self.station = _station
16           self.IAT = _IAT
17           self.ST = _ST
18           self.schedule( self, "Packet_Arrival", self.
         scheduler.time + rnd.expovariate(1.0/self.IAT) )
19
```

```python
20    def evHandler(self, ev):
21        # Handle arrival event
22        pkt = Packet( rnd.expovariate(1.0/self.ST) )
23        pkt.Arrival_Time = self.scheduler.time
24        self.schedule(self.station, pkt, self.scheduler.time
      )
25        # Schedule next packet arrival
26        self.schedule( self, "Packet_Arrival", self.
      scheduler.time + rnd.expovariate(1.0/self.IAT) )
27
28
29 class Packet:
30    def __init__(self, _size):
31        self.size = _size
32        self.Arrival_Time = 0.0
33        self.Service_At = 0.0
34        self.Departure_Time = 0.0
35
36    # Total time spent in system
37    def delay(self):
38        return self.Departure_Time - self.Arrival_Time
39
40    def info(self):
41        print("Arrival_Time = %.2f, Service_At = %.2f,
      Service_Time = %.2f, Departure_Time = %.2f" % (self.
      Arrival_Time, self.Service_At, self.size, self.
      Departure_Time))
42
43
44 class Server(SimEntity):
45    busy = False
46
47    def __init__(self, _scheduler, _station, _id):
48        super(Server, self).__init__(_scheduler, _id)
```

```python
49          self.station = _station
50
51      def evHandler(self, ev):
52          global Num_Pkts, Total_Delay
53
54          if isinstance(ev.type, Packet):
55              pkt = ev.type
56              self.busy = True
57              pkt.Service_At = self.scheduler.time
58              pkt.Departure_Time = self.scheduler.time + pkt.
    size
59              #pkt.info()
60              Num_Pkts = Num_Pkts + 1
61              Total_Delay = Total_Delay + pkt.delay()
62              self.schedule(self, "End_of_Service", self.
    scheduler.time + pkt.size)
63          elif ev.type == "End_of_Service":
64              self.busy = False
65              self.schedule(self.station, "Server_Available",
    self.scheduler.time)
66          else:
67              print("Server is supposed to receive packets!")
68
69      def isBusy(self):
70          return self.busy
71
72
73  class Station(SimEntity):
74
75      def __init__(self, _scheduler, _id):
76          super(Station, self).__init__(_scheduler, _id)
77          self.Q = queue.Queue()
78          self.S = Server(_scheduler, self, _id)
79
```

```python
80      def evHandler(self, ev):
81          # Handle arriving packet
82          if isinstance(ev.type, Packet):
83              pkt = ev.type
84              if not self.S.isBusy():
85                  self.schedule(self.S, pkt, self.scheduler.
    time)
86              else:
87                  self.Q.put(pkt)
88          elif ev.type == "Server_Available":
89              if not self.Q.empty():
90                  pkt = self.Q.get()
91                  self.schedule(self.S, pkt, self.scheduler.
    time)
92          else:
93              print("Station is supposed to receive packets
    only!")
94
95  Num_Pkts = 0.0
96  Total_Delay = 0.0
97
98  scheduler = Scheduler(100000)
99  station = Station(scheduler, 1)
100 src = TrafficGen(scheduler, station, 2, 3.33, 2.5)
101 scheduler.run()
102
103 print("Avg Delay = %.2f" % (Total_Delay / Num_Pkts))
```

Listing B.9
State.

```python
1   class State:
```

```
2    def action(self):
3        pass
4
5    def next(self, event):
6        pass
```

**Listing B.10
State Machine.**

```
1   # http://python-3-patterns-idioms-test.readthedocs.org/en/
        latest/StateMachine.html#the-table
2
3
4   class StateMachine:
5       def __init__(self, initialState):
6           self.currentState = initialState
7           self.currentState.action()
8
9       # Make transition
10      def applyEvent(self, event):
11          self.currentState = self.currentState.next(event)
12          self.currentState.action()
```

**Listing B.11
Simple Protocol.**

```
1   from state import State
2   from stateMachine import StateMachine
3   from event import Event
4
5
```

```
6    class Bad(State):
7        def __init__(self):
8            super(Bad, self).__init__()
9
10       def action(self):
11           print("Bad State")
12
13       def next(self, event):
14           if event.type == "B":
15               return self
16           else:
17               return Good()
18
19
20   class Good(State):
21       def __init__(self):
22           super(Good, self).__init__()
23
24       def action(self):
25           print("Good State")
26
27       def next(self, event):
28           if event.type == "G":
29               return self
30           else:
31               return Bad()
32
33
34   class Protocol(StateMachine):
35       def __init__(self, _initialState):
36           super(Protocol, self).__init__(_initialState)
37
38
39   p = Protocol(Bad())
```

```
40  p.applyEvent(Event(None, None, "G", None))
41  p.applyEvent(Event(None, None, "G", None))
42  p.applyEvent(Event(None, None, "B", None))
```

The Chi-Squared Table

$k-1$	$\chi^2_{0.005}$	$\chi^2_{0.010}$	$\chi^2_{0.025}$	$\chi^2_{0.050}$	$\chi^2_{0.100}$	$\chi^2_{0.900}$	$\chi^2_{0.950}$	$\chi^2_{0.975}$	$\chi^2_{0.990}$	$\chi^2_{0.99}$
1	0.000	0.000	0.001	0.004	0.016	2.706	3.841	5.024	6.635	7.879
2	0.010	0.020	0.051	0.103	0.211	4.605	5.991	7.378	9.210	10.59
3	0.072	0.115	0.216	0.352	0.584	6.251	7.815	9.348	11.345	12.83
4	0.207	0.297	0.484	0.711	1.064	7.779	9.488	11.143	13.277	14.86
5	0.412	0.554	0.831	1.145	1.610	9.236	11.070	12.833	15.086	16.75
6	0.676	0.872	1.237	1.635	2.204	10.645	12.592	14.449	16.812	18.54
7	0.989	1.239	1.690	2.167	2.833	12.017	14.067	16.013	18.475	20.278
8	1.344	1.646	2.180	2.733	3.490	13.362	15.507	17.535	20.090	21.95
9	1.735	2.088	2.700	3.325	4.168	14.684	16.919	19.023	21.666	23.58
10	2.156	2.558	3.247	3.940	4.865	15.987	18.307	20.483	23.209	25.18
11	2.603	3.053	3.816	4.575	5.578	17.275	19.675	21.920	24.725	26.75
12	3.074	3.571	4.404	5.226	6.304	18.549	21.026	23.337	26.217	28.30
13	3.565	4.107	5.009	5.892	7.042	19.812	22.362	24.736	27.688	29.81
14	4.075	4.660	5.629	6.571	7.790	21.064	23.685	26.119	29.141	31.31
15	4.601	5.229	6.262	7.261	8.547	22.307	24.996	27.488	30.578	32.80
16	5.142	5.812	6.908	7.962	9.312	23.542	26.296	28.845	32.000	34.26
17	5.697	6.408	7.564	8.672	10.085	24.769	27.587	30.191	33.409	35.71
18	6.265	7.015	8.231	9.390	10.865	25.989	28.869	31.526	34.805	37.15
19	6.844	7.633	8.907	10.117	11.651	27.204	30.144	32.852	36.191	38.58
20	7.434	8.260	9.591	10.851	12.443	28.412	31.410	34.170	37.566	39.99
21	8.034	8.897	10.283	11.591	13.240	29.615	32.671	35.479	38.932	41.40
22	8.643	9.542	10.982	12.338	14.041	30.813	33.924	36.781	40.289	42.79
23	9.260	10.196	11.689	13.091	14.848	32.007	35.172	38.076	41.638	44.18
24	9.886	10.856	12.401	13.848	15.659	33.196	36.415	39.364	42.980	45.55
25	10.520	11.524	13.120	14.611	16.473	34.382	37.652	40.646	44.314	46.92
26	11.160	12.198	13.844	15.379	17.292	35.563	38.885	41.923	45.642	48.29
27	11.808	12.879	14.573	16.151	18.114	36.741	40.113	43.195	46.963	49.64
28	12.461	13.565	15.308	16.928	18.939	37.916	41.337	44.461	48.278	50.99
29	13.121	14.256	16.047	17.708	19.768	39.087	42.557	45.722	49.588	52.33
30	13.787	14.953	16.791	18.493	20.599	40.256	43.773	46.979	50.892	53.67
40	20.707	22.164	24.433	26.509	29.051	51.805	55.758	59.342	63.691	66.76
50	27.991	29.707	32.357	34.764	37.689	63.167	67.505	71.420	76.154	79.49
60	35.534	37.485	40.482	43.188	46.459	74.397	79.082	83.298	88.379	91.952
70	43.275	45.442	48.758	51.739	55.329	85.527	90.531	95.023	100.425	104.21
80	51.172	53.540	57.153	60.391	64.278	96.578	101.879	106.629	112.329	116.32
90	59.196	61.754	65.647	69.126	73.291	107.565	113.145	118.136	124.116	128.29
100	67.328	70.065	74.222	77.929	82.358	118.498	124.342	129.561	135.807	140.16

The *t*-Distribution Table

$n-1$	$1-\alpha = 0.80$ $\frac{\alpha}{2} = 0.10$	$1-\alpha = 0.90$ $\frac{\alpha}{2} = 0.05$	$1-\alpha = 0.95$ $\frac{\alpha}{2} = 0.025$	$1-\alpha = 0.98$ $\frac{\alpha}{2} = 0.01$	$1-\alpha = 0.99$ $\frac{\alpha}{2} = 0.005$
1	3.078	6.314	12.706	31.821	63.657
2	1.886	2.920	4.303	6.965	9.925
3	1.638	2.353	3.182	4.541	5.841
4	1.533	2.132	2.776	3.747	4.604
5	1.476	2.015	2.571	3.365	4.032
6	1.440	1.943	2.447	3.143	3.707
7	1.415	1 .895	2.365	2.998	3.500
8	1.397	1.860	2.306	2.896	3.355
9	1.383	1.833	2.262	2.821	3.250
10	1.372	1.812	2.228	2.764	3.169
11	1.363	1.796	2.201	2.718	3.106
12	1.356	1.782	2.179	2.681	3.054
13	1.350	1.771	2.160	2.650	3.012
14	1.345	1.761	2.145	2.625	2.977
15	1.341	1.753	2.132	2.602	2.947
16	1.337	1.746	2.120	2.584	2.921
17	1.333	1.740	2.110	2.567	2.898
18	1.330	1.734	2.101	2.552	2.878
19	1.328	1.729	2.093	2.540	2.861
20	1.325	1.725	2.086	2.528	2.845
21	1.323	1.721	2.080	2.518	2.831
22	1.321	1 .717	2.074	2.508	2.819
23	1.320	1.714	2.069	2.500	2.807
24	1.318	1.711	2.064	2.492	2.797
25	1.316	1.708	2.060	2.485	2.878
26	1.315	1.706	2.056	2.479	2.779
27	1.314	1.703	2.052	2.473	2.771
28	1.313	1.701	2.048	2.467	2.763
29	1.311	1.699	2.045	2.462	2.756
≥ 30	1.282	1.645	1.960	2.327	2.575

Bibliography

[1] G. E. P. Box and Mervin E. Muller. A Note on the Generation of Random Normal Deviates. *The Annals of Mathematical Statistics*, 29(2):610–611, 1958.

[2] Mohsen Guizani et al. *Network Modeling and Simulation: A Practical Perspective*. Wiley, Hoboken, New Jersey, 2010.

[3] James E. Gentle. *Random Number Generation and Monte Carlo Methods*. Springer, New York City, 1998.

[4] J. M. Hammersley and K. W. Morton. A new Monte Carlo technique: Antithetic variates. *Mathematical Proceedings of the Cambridge Philosophical Society*, 52(3):449–475, 1956.

[5] Stewart Hoover. *Simulation: A Problem-Solving Approach*. Addison-Wesley, Boston, 1990.

[6] Schruben Lee. Simulation Modeling with Event Graphs. *Communications of the ACM*, 26(11):957–963, 1983.

[7] J. D. C. Little. A Proof of the Queueing Formula $L = \lambda W$. *Operations Research*, 9(3):383–387, 1961.

[8] M. H. MacDougall. *Simulating Computer Systems: Techniques and Tools*. The MIT Press, Cambridge, Massachusetts, 1987.

[9] Christof Paar and Jan Pelzl. *Understanding Cryptography*. Springer, New York City, 2010.

[10] S.K. Park and K.W. Miller. Random Number Generators: Good Ones are Hard to Find. *Communications of the ACM*, 31(10):1192–1201, 1988.

[11] P.D. Welch. *The Computer Performance Modeling Handbook*. Academic Press Cambridge, Massachusetts, 1983.

Index